Metric, [illegible] &

Quasicrystals

Antony J. Bourdillon

UHRL, PO Box 700001, San Jose CA 75170, consultant to
Institute for Superconductivity and Electronic Materials
University of Wollongong

AuthorHouse™
1663 Liberty drive
Bloomington, IN 47403
www.authorhouse.com
Phoone: 1-800-839-8640

Published by AuthorHouse 09/18/2012

ISBN: 978-1-4772-4786-0 (sc)
ISBN: 978-1-4772-4785-3 (e)

And by UHRL
ISBN: 978-0-9789-8393-2;

Front cover shows an icosahedron at right with an embedded golden triad at left, defining Cartesian axes.

Logarithmically periodic solids bibliography

Logarithmically Periodic Solids, Nova Science, 2011, ISBN 978-1-61122-977-6
Quasicrystals' 2D tiles in 3D superclusters, UHRL, 2010, ISBN: 978-0-9789839-2-5
Quasicrystals – and quasi drivers, UHRL, 2009, ISBN 978-0-9789-8391-8
Logarithmically periodic solids – properties, evidence and uncertainties, in *Quasicrystals: Types, Systems, and Techniques,* Nova Science, 2010, ISBN 978-1-61761-123-0

A.J.Bourdillon, Nearly free-electron energy-bands in a logarithmically periodic solid, *Solid State Comm.* **149** 1221-5 (2009).
A.J.Bourdillon, Indexed scattering powers in a logarithmically periodic solid, *International Journal of Condensed Matter, Advanced Materials, and Superconductivity Research* (2010)
A.J.Bourdillon, Fine Line Structure in Convergent Beam Electron Diffraction in Icosahedral Al_6Mn, *Phil. Mag. Lett.* **55** 21-26 (1987)

http://www.quasicrystalmetric.us The metric for the quasi-Bragg law measured for the first time
http://www.quasicrystal.us The Quasi-Bragg law and metric - from geometric scatterers in three dimensional space.
http://www.youtube.com/watch?v=0LDS0sQQvpk : the Quasi Bragg law and metric
http://www.youtube.com : quasicrystals, logarithmically periodic
http://www.youtube.com : quasicrystals, here are the atoms
http://www.UHRL.net, the structure of quasicrystals
http://www.quasicrystaltiling.us two-dimensional tiles in three-dimensional space – edge sharing tiles in logarithmically periodic solids.

Contents

1. Introduction 1
2. The metric for the quasi-Bragg law measured for the first time 9
3. Evaluation 65
4. Index 110
5. Reference 113

Chapter 1

Introduction

Quasicrystals have intrigued a few 'scientists' for many years. After 80 summers of practiced crystallography, these materials were discovered to have 5-fold symmetries. This is impossible in crystals, where space filling of unit cells restricts them to 14 types of Bravais lattice (cubic, tetrahedral, orthorhombic etc.), none of them having that symmetry. The purpose of this book is to unconfuse after 30 subsequent winters. It shows that quasicrystals are about as ordinary as window glass. They require no special understanding.

Structure is important in the study of materials of all types. It determines most of the physical properties that we examine in science. As always, observation and measurement are critical. In this book we place special emphasis on the metric, which has had a fundamental role in modern science. Since Einstein, it has related covariant to contravariant components of invariant vectors. In crystallography, it relates measurements in the diffraction pattern to locations of atoms in real space. In this book, the first measurement of the metric in quasicrystals is described.

The book is arranged firstly to describe experimental and computational results and secondly to aid the reader in assessment by including debates with journal referees. The debate has to occur, though inconsistncies between anonymous referees, destroys their credibility. Starting with the particular, what matters when a paper is refused by some while the theory is true? In a scientific revolution the reviewing system does not work because most referees are competitors. This is not to say that all referees have rejected our theory. It is accepted by a reviewer in *Acta Crystallographica A* (chapter 3) and in *Solid State*

Communications it has been embraced enthusiastically, "This is new to me and interesting." Two good reviews outweigh a thousand that are illogical. All have been analyzed, some in this book, others previously. The reader may judge for himself. He will find that no argument against the theory has been given that is not either false in fact or fallacious in logic. However the reader will surely wonder whether such a reviewing process is not wasteful of voluntary effort. When neither scientists nor editors follow the demands of logic, medieval orthodoxy will continue to reign.

The discovery of quasicrystals is sometimes called a revolution in science. The International Union of Crystallography (IUCr) even chose to redefine crystals:

> "A material is a crystal if it has essentially a sharp diffraction pattern. The word essentially means that most of the intensity of the diffraction is concentrated in relatively sharp Bragg peaks, besides the always present diffuse scattering. In all cases, the positions of the diffraction peaks can be expressed by
>
> $$\mathbf{H} = \sum_{i=1}^{n} h_i \mathbf{a}_i^* \quad (n \geq 3)$$
>
> Here $\mathbf{a}_i^*$ and h_i are the basis vectors of the reciprocal lattice and integer coefficients respectively and the number n is the minimum for which the positions of the peaks can be described with integer coefficient h_i.
>
> "The conventional crystals are a special class, though very large, for which $n = 3$." [i]

The first of many myths is that diffraction in quasicrystals is Bragg diffraction. In the weakest possible sense, perhaps it is, but not in any sense that Bragg described. The weak sense is that the

---

[i] *Acta Cryst.* (1992), **A48**, 928 where the definition of a crystal appears in the Terms of reference of the IUCr commission on aperiodic crystals.

diffraction is due to a ragged three dimensional grating where diffracting planes of atoms are variably spaced. In this very weak sense, the diffracting planes are normal to the scattering vector, a feature that we use in our simplified indexation, or naming, of the pattern. In the sense of Bragg's law, the diffraction from quasi crystals is not Bragg diffraction. All orders are forbidden except the second, and that only with modification. This law describes diffraction due to a lattice of atoms onto a lattice in reciprocal space. In quasicrystal diffraction, the pattern in reciprocal space is not a regular lattice. The angular spacing between diffracted beams is not even a Fibonacci series as commonly claimed, because the series ratio, F_{n+1}/F_n is not a varying number tending to the golden section τ at large n; but is the constant τ. It is therefore a geometric series; far from the arithmetic series prescribed by Bragg's law. (Reader can find an example of the quasicrystalline pattern in chapter 2 [ii].) In quasicrystals, the lattice is logarithmic[iii]. It is contradictory[iv] to claim the diffraction is both Fibonacci and Bragg. There are other differences too: among them the quasi-Bragg law has a special metric that is described in this book. This metric is measured and explained for the first time. If IUCr were to need to include quasicrystals in their definition, then it would have to redefine crystals again. It is no use saying we must do as they do – not while we use a valid optional convention of indexation that is consistent with all measurement. With the proper law it is possible to measure atomic positions from the diffraction pattern as is typical in crystallography, and we find these are consistent with transmission electron microscope (TEM) imaging exactly, and also with known atomic sizes, known that is for the first time.

[ii] Or see an elementary, comparative illustration in *Quasicrystals and quasi drivers, UHRL 2009 pp. 30-31* ISBN 978-1-4389-5589-6

[iii] Bourdillon, A. J., Nearly free electron band structures in a logarithmically periodic solid, Sol. State Comm. **2009**, 149, 1221-1225.

[iv] The truth of the contradiction is independent of the number of 'scientists' that have made it.

This book is written so that the reader may independently evaluate the great divide. X-ray crystallographers and electron microscopists use indexation differently. Their methods of indexation are correspondingly different but equally valid according to Bragg's law. The X-rays have typical wavelengths about 0.1 nm, short of some of the atomic interplanar spacings. In transmission electron microscopy (TEM), electron wavelengths are 30 times shorter. Scattering angles are correspondingly smaller. The restriction, imposed by the Ewald sphere construction in X-ray diffraction (XRD), is relaxed in TEM where the deviation parameter takes complementary significance. In a vertical mount TEM all Bragg planes contain the vertical axis, and all their normals are horizontal. TEM is content to name the beams on a plane, without concern for full circle diffractometry. The whole pattern is indexed from a few beams around each axis[v] This is the method used here and we show that three dimensions are sufficient and easily manageable. It is unnecessary to index every axis separately because of symmetry. The icosahedral structure has few axes. The three main axes are, on Cartesian coordinate axes, the $[0\bar{1}\tau]$ for the 5-fold axis; the [111] for the 3-fold axis; and the [100] for the 2-fold axis. These are illustrated in the following chapter. Allowing for sign change and for permutations, the first has 12 variations that match the 12 corners of the icosahedron. The 3-fold [111] axis allows 8 permutations, which is less than the 20 triangular faces on the icosahedron. There are therefore a further 12 variants of the axial indices and their patterns could be indexed either by unitary transformations, based on the [111] axial pattern, or simply by the same method repeated on the easily identified axes. The 2-fold [100] axis has 6 index permutations. This is a minority of the 30 sides on the icosahedron. Each side has at its center a 2-fold axis, though the one we use is alone sufficient to index the pattern on one plane, in order to confirm the model and to calculate the metric. The axial patterns are all indexed in 3

---

[v] Hirsch, P.; Howie, A.; Nicholson, R.B.; Pashley, D.W.; Whelan, M.J., *Electron Microscopy of thin films*, 2nd ed. 1977, Krieger, NY, appendix 5.

dimensions and this is of significant computational convenience. It would be unscientific to offer a hostage to error by adopting a convention of unnecessary complexity and unexplored metric. Our indexation of all beams on all axial patterns is complete. The full stereographic projection of axes is indexed easily in three dimensions (ch. 2, appendix 2). Any comprehending reader will find our method the cleanest and simplest one ever described. It yielded the metric.

There is moreover an additional reason for indexing the diffraction patterns *ab initio*, and that is to assign the quasi lattice parameter that has been measured by several researchers. Our measurement and method have been previously published[vi] and so are only referenced in the following chapters. Happily the assignment is unambiguous and has been independently adopted by other workers[vii,viii]. In summary, the quasi lattice parameter is 0.206 nm, which with the index $(2/\tau,0,0)$, corresponds to the subcluster side of 0.269 nm with metric applied, and to the subcluster length of 0.435 nm. The indexation has a natural connotation. It corresponds to an interplanar spacing half the length of the unit cell, which is also half the length of separation between the centers of two neighboring cells. It is far from surprising that reflection from corresponding planes produces the strongest beams in diffraction. This is confirmed by simulation. The assignment is therefore consistently circular.

Most, if not all the reviews analyzed in this book, are written by members of the X-ray or mathematics communities. That is why they have not understood our very simple and reliable method of indexation, effective as it is in the measurement of the metric, and consistent with the quasi-lattice parameter. However, so long as the indexation is proper, it does not matter which nomenclature is

---

[vi] Ibid. appendix A

[vii] Tsai, A.P., Icosahedral clusters,, icosahedral order and stability of quasicrystals – a view of metallurgy, *Sci.. Technol. Adv. Mat., 9 1-20 (2008)*

[viii] Akakura, H, Gomez, C.P., Yamamota, A., De Boisseau, M., Tsai, A.P., *Nature Materials* **6** 58-63 (2006)

selected because the 'structure factors' that we calculate are invariant.

The core of chapter 2 has been written in two forms. For those who know their way about, a condensed form was submitted to *Acta Crystallographica A* and is published on www.quasicrystal.us It became clear that referees do not read the references given, so the paper was expanded in response to an invited review request from *Materials* for a special issue (on quasicrystals - and myth as it turns out). Chapter 2 in this book contains material that features important information, including the 5-fold diffraction pattern in i-Al_6Mn. The web page therefore concentrates the material for readers familiar with the topic; while chapter 2 provides easy access to relevant supporting information. This is admittedly limited by the restraints of copyright. The reader may choose his medicine neat, or embellished in a cocktail. Are the data effective? No two reviewers made the same criticism and we have to speculate on reasons for their inconsistencies. Try the big picture and then hone the circumstances of the individual reviewers. Most scientists are called to review, some more than others; the wonder is that reviewers mistake so often their a priori opinions as privileged over fact. Their opinions are not subjected to the normal rules of science and therefore match corresponding credulity. The outcome is positive: *this is the first measurement of the metric.*

Science is a 2300 year old enterprise. Aristotle's Physics is supported by his Organon, six works on logic. His formal logic and informal logic were endorsed by the philosopher John Locke [ix] during the enlightenment. Since then science has become an industry so that what he thought to be the paradigm of knowledge has become a network of special and political interests. Meanwhile, the principles of logic are observed by neglect by both reviewers and editors. There is a chapter of examples. The system of anonymous refereeing is a license to property theft and a cloak for irresponsibility, usually shielded by unresponsive editors - not

---

[ix] See *The Cambridge Dictionary of Philosophy,* ed. R.Audi, CUP, 1999

entirely, as the reader will find. False logic led to the silencing of Galileo and evidently continues its action. In particular, *ad populum* arguments are as common in 'science' as they are fallacious and myth building. There are structural problems that drive the illogicality. Among them, why is publicly funded research not publicly available in this electronic age? The implication is that journals distort funding. However, at a deeper level they perform disservice: typically their guides for referees ask not, "Is a submission true and if not why not"; instead, they ask a group of softer questions equivalent to, "Does it confirm existing myth?"[x] Then they support mediocrity and continuing confusion.

Returning to this book, which does not describe the uses of quasicrystals, it is not surprising that their unusual physical and electronic structures accompany characteristic properties. Their Uses are described elsewhere in various places[xi].

A comment on references: in order to differentiate citations, different calling methods are used in the three chapters. To avoid confusing references, chapters 1 and 3 use the footnote method. In chapter 2, endnotes are numbered in a manner typical of modern journal articles. This way, Reader won't mix them up. Notice that some of the arguments in chapter 3 are incorporated back into chapter 2, embellished over the condensed website already cited. In chapter 2, work is referenced only that has been used in developing the finding. Extraneous work is not referenced, so that unnecessary controversy is avoided. If at the end, Reader does not find unscientific practice anywhere, then this book will have failed. Against knowledge, myth is cancerous.

But steady on. This book is written for two purposes:; firstly for the student of quasicrystal structures, it proves that Bragg diffraction cannot be Fibonacci and there is an alternative, proper and classical way. Secondly, for the student of scientific method, the book supports the thesis that, in revolutions, the refereeing

---

[x] *E.g.* http://iopscience.org/referee-guide, /HOW TO WRITE A REFEREEE REPORT.

[xi] *E.g. Quasicrystals, Types, Systems and Techniques,* Ed. Beth E.Puckerman, Nova, 2011

system does not work for competitive reasons that deny logic. The former student can skip sections 7 and 8 in chapter 2 without great loss, and sections 3.3 to 3.6 in chapter 3. The response in 3.2.1 gives the most significant part of the evaluation.

Chapter 2

The metric for the quasi-Bragg law measured for the first time

Invited review for Materials

Forenote:-with its origins in *Sol. State Comm.* **2009**, 149, 1221;
-with www.quasicrystal.us , a concise development and debate,
-and with the graphic introduction on
http://www.youtube.com/watch?v=0LDS0sQQvpk ;
This is a further expansion designed for easier assimilation, but see the 'steady on' comment on page 8.

Contents

1. Introduction to the topic 11
2. Icosahedral quasicrystals of the first type 12
   2.1 Requirements for the Bragg Law 13
   2.2 Structural concept 16
   2.3 Interplanar spacings and preliminary calculations of scattering 21
3 Geometric and Fibonacci series 36
4. Detailed indexation and measurement 43
5. Beam intensity rankings 44
6. Modeled micrographs 51
7. Defects and holes 53
8. Other systems
   8.1 A hierarchic solution for icosahedral quasicrystals type II 55
   8.2 Decagonal quasicrystals and others with reduced symmetry 61
   8.3 The hierarchic method 61
9. Conclusions 62
Acknowledgment 63

Abstract: *The metric is the necessary rule that relates atomic spacings in real space with angles of diffraction in reciprocal space. Quasicrystals display diffraction orders in geometric series. They do not therefore follow Bragg's law of diffraction, where the series are in linear order on a reciprocal lattice. All Bragg orders are forbidden except the second, and that only with modification. The geometric series is a restricted form of Fibonacci series and has special properties since the sequence ratio between members is a constant τ, the golden section. The diffraction patterns are indexed in three dimensions with notational economy, with demonstrated completeness, without redundancy, and with measured consistency. Simulations on a hierarchic model show that the 'quasi lattice parameter' that is measured in the diffraction, differs from the Bragg equivalent in crystals. This is due partly to a compromise in the multiple interplanar spacings that contribute to the diffraction. The spacings provide approximately half integral coherence at short range, with near integral coherence at long range. A special metric relates the diffraction pattern to the corresponding interplanar structures. The simulations match data in several ways: diffracted beam intensities are correctly calculated; a high resolution electron micrograph is modeled in detail; and the edge sharing unit cell fits known atomic sizes. Comparisons are made between various alloys. Without the metric there is no measurement. It is invariant, independent of indexation.*

---

1. Introduction to the topic

"What is most intriguing, of course, is whether we are concerned with a material having singular structural properties because of the chemistry of Al and Mn, or whether the principles suggested by the quasicrystal concept will find more widespread application [1] ". The discovery of quasicrystals [2] in 1982 was a surprise that came after 80 years of crystallography and Bravais lattices. But the claim of scientific revolution [3,4] left a doubt about traditional chemical explanations. The solution that we suggested, proposed, and developed is typical for materials and follows an elementary requirement in science to subdue *a priori* assumptions to attentive study of data. These are not all equally telling. Among many published high resolution images, it was not those observed at the highest resolution, but those observed at 'optimum defocus', that yielded the subclusters [1] in skeletal form [5]. Meanwhile, our own early studies showed systematic planar defects [6] that caused superposed interference in convergent beam diffraction patterns. The fact of defects is not surprising in material produced by rapid solidification, methods often used to produce metallic glass. However the quasicrystal is a quasi stoichiometric alloy, so it was a short step to suppose that the quasi glasses contain cells, like the tetrahedra in fused silica, and that the cells should likewise be edge sharing. These general expectations follow the material production methods. The main contrast comes with the sharp diffraction patterns that are observed in the quasicrystals, as sharp as from crystals, as we shall see. So the orientations of the edge sharing cells must be ordered. This occurs when the cells share more than one edge each. Our break-though arrived, after some thoughtful delay, with the realization that the properties could be calculated from an ideal hierarchic structure based on an icosahedral unit cell. The cell itself satisfied the known stoichiometry and satisfied the required angular symmetries, as did the hierarchic model. Simulations of diffraction patterns [5], of electronic band structures due to logarithmic quasi-Block waves [7], of two-dimensional tiling [8], and now of the quasi-Bragg law, were natural consequences. Such properties translated to the defective structures that were found in imaged foils [5,9], and simulated.

Just as not all data are equally telling, so models and theory are not equally illuminating. The original data was electron microscopic, and X-ray crystallographers have often failed to understand them. The first was the late Linus Pauling [10] who claimed that the 'icosahedral' appearance was due to twinning. Others have taken short cuts by thinking of a 'Fibonacci' sequence in the diffraction patterns as a property of Bragg diffraction [3]. Complicated, multidimensional indexation has been devised [11,12]. Moreover, several 'simulations' of diffraction all failed to notice that the 2-fold patterns in the original data [2] are inconsistent. The patterns should be mutually perpendicular though they are shown parallel on respective n-fold axes (5-2-5 or 5-3-2-3-5). Short cuts in research make shorter reviews, where unnecessary dispute is unprofessional. Various alloys and structures have been developed since that early discovery. In this paper we spread the net to compare other alloys with the original discovery. Some early data is repeated in the present work in order to make it self sufficient. Brief enquiries are directed at chemical composition, stoichiometry, atomic size, processing, and non-icosahedral quasicrystals. Earlier workers have asked, "Where are the atoms?"[13]. We find that knowing where they are is part of "Why are they there?" A graphic introduction is given elsewhere [5] and a debate is recorded.

The metric that we have discovered and measured for the first time is invariant to the method of indexation used. Indexation of diffraction patterns depends on conventions. It is computationally necessary to choose the most economical method since it is least likely to lead to error, especially in large calculations. In modern science the metric has come to have a fundamental role, where in crystallography it is of corresponding importance in measuring atomic coordinates in real space from diffraction patterns in momentum space.

2. Icosahedral quasicrystals of the first type

Many structural models have been proposed for quasicrystalline alloys [*e.g.*14]. We start with symmetry, and then admit the constraints of stoichiometry, and atomic size.

The search includes a unit cell but, as in fused silica, does not require face sharing - most notably in metastable, rapidly quenched material which appears partly glassy. Options are not wide, and we consider first the most obvious and simplest.

2.1. Requirements for a quasi-Bragg law

The diffraction pattern of quasicrystals is of extraordinary form and it is necessary first to describe it accurately. In crystals, the Bragg law describes a lattice of atoms by mapping them onto a reciprocal lattice. Such lattices are observed most easily in high energy electron diffraction [15]. In quasicrystals the reciprocal lattice is in geometric space which is a restricted form of 'Fibonacci space'. In a Fibonacci series, the sequence can be written: *a; b; a+b; a+2b; 2a+3b; 3a+5b;...*, where each term is the sum of the two preceding terms, and where a, b are real numbers. As the sequence progresses, the ratio between terms, $F_{n+1}/F_n \rightarrow \tau \equiv (1+\sqrt{5})/2$, the golden section [16]. However, when $a=1$ and $b=\tau$, the series can be written geometrically, $1;\tau;\tau^2;\tau^3;\tau^4;...$ With this representation the quasi-Bragg formula maps lattice-like onto a logarithmic reciprocal space. This is essential to the correct description of the law of diffraction.

However in this case, the mapping is not as straight forward as it was for Bragg. Before reaching for the formula, here is a description of the process, given as a heuristic example:

> "Naturally, as I was constructing the model, I was thinking about how to simulate the diffraction pattern. Understanding the relation of Bragg's law to structure factors, I was able to arrive at the pattern by what I at first expected would be straight-forward means. The version of Bragg's law was the same as (previously) except that d_H was in first order *i.e.* with $d=a/h$, a being the edge length of the unit cell, and h the index, derived unambiguously by prior means... To my surprise the structure factors were all low and haphazard. I knew what values to expect because it depends on the number of scattering atoms. By scanning the 'Bragg angle' reflection, I found the quasi Bragg angle, θ' and corresponding quasi interplanar spacing d'. I then

set about checking that the diffraction pattern was indeed logarithmic (or geometric or restricted Fibonacci) and that the second order Bragg (n=2) was zero because higher orders are not (with exceptions) observed in the diffraction patterns. To my consternation the structure factors increased by orders of magnitude in second order and it took me a long time to work out why. I found moreover that *d'/d* was the same for all peaks. I did simulations on imaginary systems that are different in long and short range and found that the reason for the compromise spacing effect (or *d' /d* or the metric or whatever) is due to the contrast in properties of the geometric series in long and short range. I could predict and measure *d'* directly and consistently relate it to the structure."

To arrive at the quasi-Bragg law, a model was needed. Partly because of this logarithmic lattice, we called it the logarithmically periodic solid (LPS)[17]. This is an ideal hierarchic structure that is built from edge-sharing icosahedral subclusters. The structure is infinitely extensive, uniquely aligned and uniquely icosahedral (figure 1). Simulations showed that the alignment provides the sharp diffraction pattern and the structure yields diffracted intensities that match experimental values [5]. The quasi-Bragg law was written [16]:

$$\lambda = d'\tau^{-m}\sin(\theta') \tag{1}$$

where m is the logarithmic order, θ' is the quasi-Bragg angle, and $d'=d/c_s$ *i.e.* the classical interplanar spacing corrected by the metric c_s. This metric was given by the difference from the expected angle at which diffraction was classically derived in structure factor calculations. Analysis of the geometric series will show why the quasi-Bragg law is in second Bragg order, causing the factor 2 in the Bragg law to cancel. Diffraction pattern intensities simulated remarkably well as will be described in section 5. In the geometric space, the axial patterns are unambiguously and completely indexed in three dimensions without redundancy. The notational economy is due to the application of the quasi-Bragg law on the basis of a tetragonal subgroup (T_d, see cover page) of the icosahedral point group symmetry. The economy is especially useful both in electron microscopy where selected choice of axial indexation bears no disadvantage, and is also useful in the computational methods and novel measurements described below. An alternative six dimensional representation is used elsewhere [3,4], with separate indices measuring units of 1 and units of τ.

Figure 1a. Icosahedral cluster made from 12 edge sharing icosahedral subclusters, including principal axes.

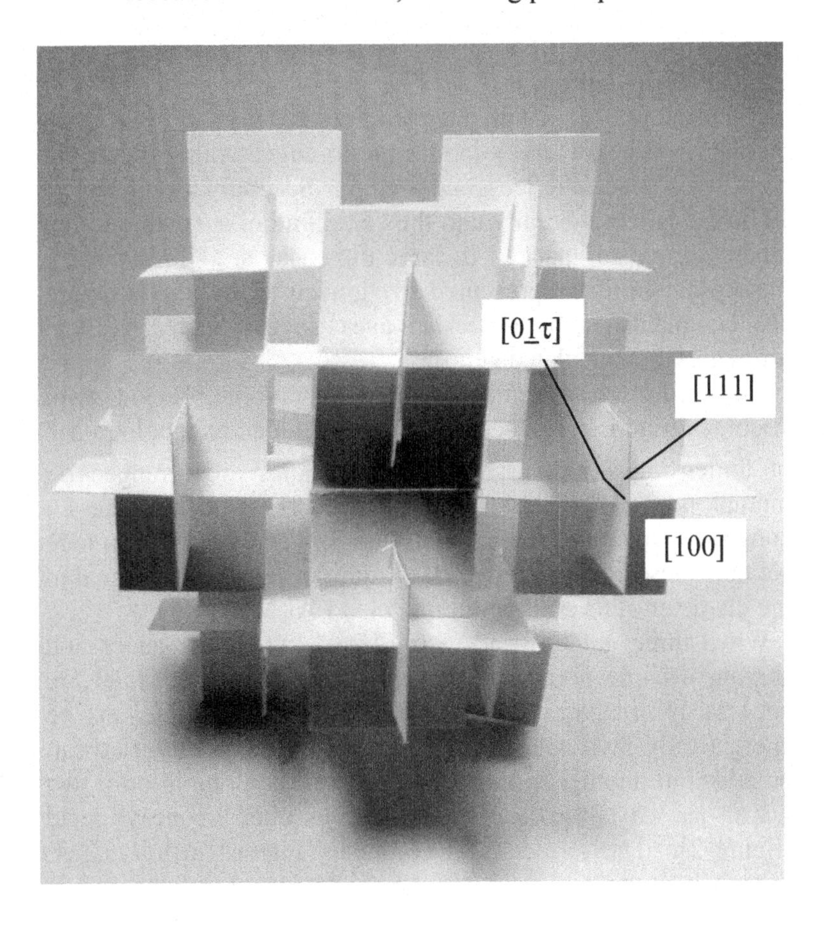

Figure 1b. Atomic model of subcluster.

2.2 Structural concept

The concept begins with a triple intuition. Firstly, when three regular icosahedra are joined edge-to-edge, with one 'triple point' of contact, the icosahedra align. Secondly a unit cell of 13 atoms is icosahedral when the central atom is small and the others are closely packed in three dimensions. Thirdly, when the icosahedron is represented by 'golden triads' the structure can be quantified on Cartesian axes, so that many physical properties are calculated simply.

The third intuition can be used to illustrate the previous two. A golden triad is constructed from three golden rectangles, each of dimensions $1\times\tau$. The three rectangles are arranged on mutual normals. The triad has 12 corners that locate 20 congruent triangular faces that in turn join 30 edges. From here we take these equal edges to be our unit of length. It is equal to the diameter of an Al atom in modeled i-Al_6Mn.

When three golden triads are aligned by joining edges and sharing a triple point, the centers of the resulting triplet are necessarily planar. Supposing the triple point to be energetically favorable, consider how a fourth golden triad can be added at another triple point. This can be done in only two ways: one way forms a planar quad; the other a concave quad (figure 2). The former is of secondary interest in this paper, though it is most likely instrumental in planar defects and in other varieties of quasicrystal, such as those having linear periodicity in one direction. The concave quad is more interesting: four overlapping quads form an icosahedral cluster made from 12 icosahedral golden triads (figure 1). 12 clusters form an icosahedral supercluster, and the series repeats indefinitely. The resulting superclusters are aligned, uniquely icosahedral and infinitely extensive. The stretching factor between orders is τ^2 .

Figure 2. Planar quad (left), and concave quad (right) form the complete set of four triadic golden rectangles with two triple points.

Three immediate conditions (figure 3) for this structure are:

-firstly, if the central atom is called a 'solute' (as in the melt before crystallization), then the ratio of the diameters of the solute atoms to the solvent atoms (here having unit length like the icosahedral edge) is $\sqrt{\tau^2+1}-1$;
-secondly the stoichiometry of solute to solvent is 1:6, after accounting for shared boundary atoms.
-thirdly the coordination of 12 about the solute atom is optimum in the metallic system.

Figure 3. The cross-section of a subcluster demonstrates atomic size requirements for binaries in the icosahedral structure. In the golden rectangle, the solute (filled circle) diameter is $\sqrt{1+\tau^2}$ -1

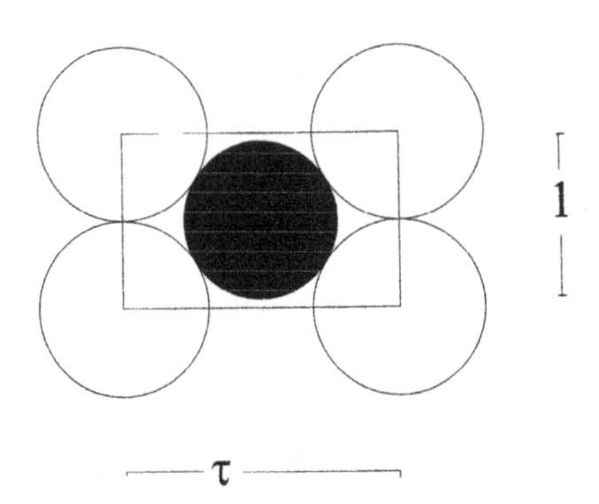

For interest, the three conditions are consistent with the original data [2] and with all type I binary i-quasicrystals. The conditions are modified in type II i-quasicrystals as will be discussed in section 8.

The unit cell is dense. It is compared, in figure 4a, with a corresponding cell in face centered cubic Al. Melt spun Al_6Mn has two phases. In the crystalline matrix phase, the smaller solute atom rattles in a (111) raft of Al. In the quasicrystal phase, three atoms move up on the raft shown in the figure, and three move down. The cell compresses around the small solute atom, taking away the rattle space. By use of the quasi-Bragg law, a density increase of 17% is measured. This seems to be the driving force for the structure.

We have called the cluster illustrated in figure 1 'logarithmically periodic' [7] for many reasons. These include the peculiar periodicity and indexation of the diffraction pattern. The diffraction pattern entails the same point group symmetry as the structure so we index for icosahedral or dodecahedral symmetry. Proceed by 'indexing' the principal axes $[0\bar{1}\tau]$ [111] and [100] as in figure 1a. Some of these numbers are irrational, but this is not surprising since the orders are all integral exponents on the base τ. The indices could therefore be re-written $[-\infty,\underline{0},1]$, $[000]$, and $[0,-\infty,-\infty]$ respectively, though it is not convenient to do so. We shall attempt to apply conventional methods, remembering meanwhile, limitations for a quasi reciprocal unit cell that is imagined intuitively and not derived from primitive lattice vectors. Notice that while it is not necessary to use more than three indices to label each electron beam reflection, our particular methods adapt well to our convention in indexation. Notice further, that uniqueness in indexation is not a requirement in crystallography. For example, electron microscopists index the hexagonal close packed structure in three dimensions [15]; X-ray crystallographers always prefer more, in this case four. [20].

Figure 4a. Two unit cells compared: at right a unit in the face centered cubic (fcc) matrix phase of Al_6Mn, where the small (scaled and filled) solute atom has room to rattle; at left, an icosahedral unit in which three atoms in the central raft move up and three down. The structure compresses around the solute atom with a density increase of 17%.

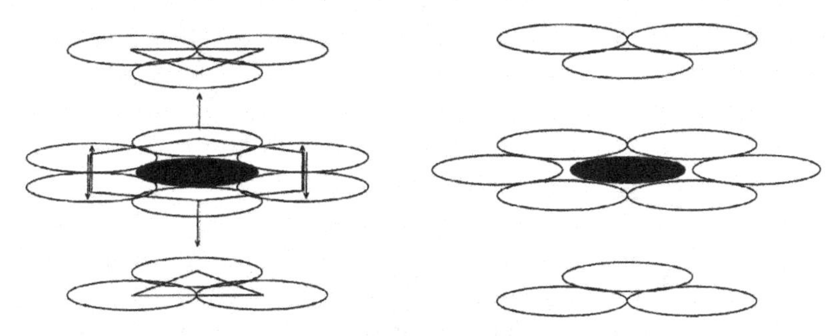

The fcc structure is compared again in the stereographic projection shown in figure 4b. Notice the identities and equivalents: $[\tau 00]$ is identically oriented with [100] in the fcc stereogram[15]; $[\tau^2\tau^2\tau^2]$ is identically oriented with [111]; and

Figure 4b. Stereographic projection of 5-fold, 3-fold and 2-fold axes in the icosahedral structure using 3-dimensional indexation. The central axis is the 5-fold $[0\bar{1}\tau]$. The outer circle projects from the point $[01\bar{\tau}]$.

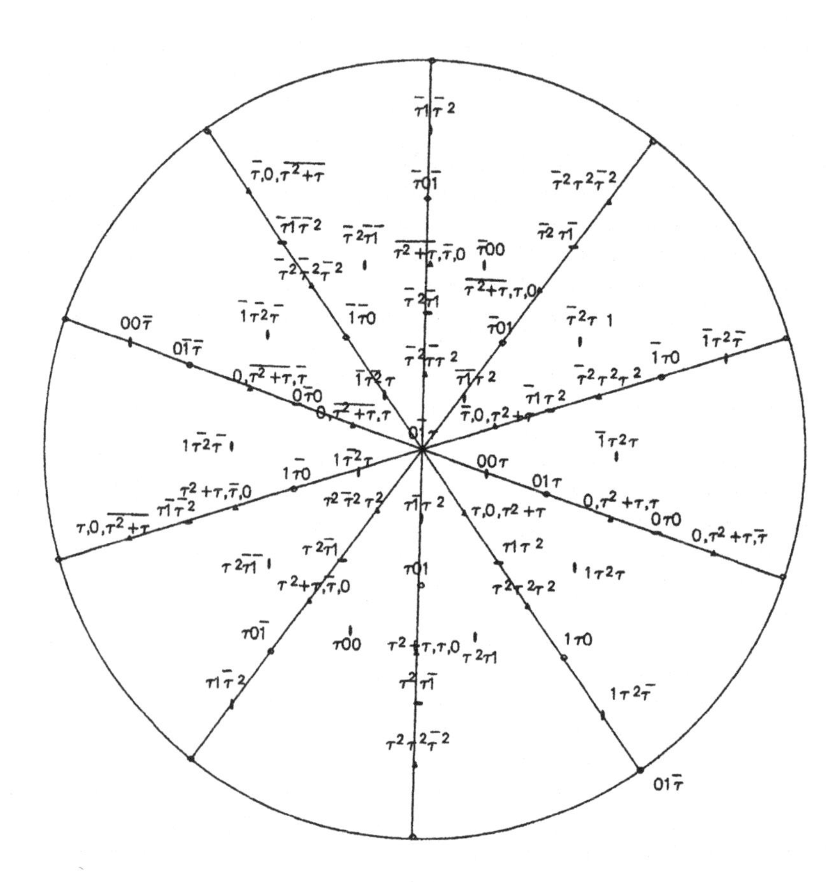

$[0\bar{1}\tau]$ is roughly equivalent to $[0\bar{2}3]$. It is easier to us the fcc notations for the two identical orientations and so we have adopted these conventions for this book.

Meanwhile, as a preliminary to indexation about the $[0\bar{1}\tau]$ axis to be discussed in the following section, an illustration is given in figure 4c which shows the locus of nearest neighbor cells on the $(0\bar{1}\tau)$ plane around a unit cell.

Figure 4c. The locus coordinates of centres of neighboring cells on the $(0\bar{1}\tau)$ 5-fold plane in the hierarchic model.

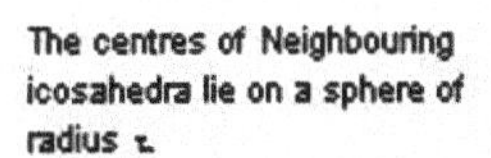

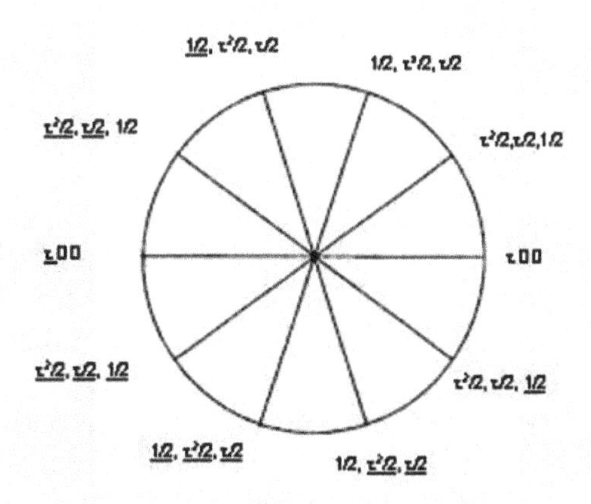

As we shall see, the locations are an example of general consistency between the indexation of the diffraction pattern and corresponding cellular and atomic sites. Notice how easy it is to index all major axes and to proceed to index, by the simple method described, corresponding third bright rings in the diffraction patterns. From these rings, complete indexation follows in a clean, simple and straight forward way. This is all easily done in three dimensions by methods that are conventional in electron microscopy and that apply, in the obvious way, to logarithmic patterns.

Before extending our indexation to diffraction patterns, it is enlightening to consider maps of atoms in the cluster. After that, we will be set to calculate the diffraction scattering due to a large number of precisely located atoms on evident reflecting planes.

2.3. Interplanar spacings and preliminary calculations of scattering

Figure 5a shows a projection onto a horizontal plane, or map, of the atoms in a supercluster order 1, oriented to the vertical 5-fold $[0\bar{1}\tau]$ axis. Filled circles represent Mn atoms; open circles

Al atoms in modeled i-Al_6Mn. With inversion symmetry in the model, the diffraction pattern will repeat by angles $2\pi/10$. We begin by finding 10 diffraction planes normal to $[0\bar{1}\tau]$ (Schoenflies D_5). They are:
$(2/\tau,0,0)$, $(1,1/\tau,1/\tau^2)$, $(1/\tau^2,1,1/\tau)$, $(-1/\tau^2,1,1/\tau)$,
$(-1,1/\tau,1/\tau^2)$, $(-2/\tau,0,0)$, $(-1,-1/\tau,-1/\tau^2)$, $(-1/\tau^2,1,1/\tau)$,
$(1/\tau^2,-1,-1/\tau)$ and $(1/\tau^2,-1,-1/\tau^2)$.
In all cases this indexation was found (appendix A.1) to correspond with identified scattering planes in figure 1. Based partly on published diffraction patterns [2,6 part copied in section 7], extend these indices radially by multiplying by τ^m, $-2\leq m\leq 3$.., where m is integral. This writes logarithmic periodicity into the indexation as will be illustrated in a later section. The series is, as before, the geometric sequence and will be discussed again below. It is obvious that the arithmetic orders in the Bragg law $n\lambda = 2d\sin(\theta)$ are, in logarithmic periodicity, converted to indices that correspond to, $\tau^m\lambda = d\sin(\theta)$. However, with such a radical change, it is no longer clear that d is a recognizable interplanar spacing, and a debate is needed about how a modified law applies when unit cells are not arithmetically periodic, as in crystals having a Bravais lattice. In the following text we will discover another change when we put the Bragg order n=2. Simulations of atomic scattering from extensive solids will answer the uncertainty. Meanwhile intervening points in the reciprocal lattice plane of the diffraction pattern can be completely filled by vector arithmetic on the indexed logarithmic radials. The method is typical in electron microscopy. It is obvious that, with this model, the more complicated 6D notation [11] and use of hyperspace are not needed. A feature of the 5-fold pattern is that complete indexation can be made by projection of the 'third bright ring' (illustrated in section 7). The equivalent bright ring, along with the geometric series is used to index the 3-fold pattern. This pattern (Schoenflies D_3) can be indexed in a similar way. Six indices normal to [111] are:
$(-1,1/\tau,1/\tau^2)$, $(-1/\tau^2,1,-1/\tau)$, $(1/\tau,1/\tau^2,-1)$, $(1,-1/\tau,-1/\tau^2)$,
$(1/\tau^2,-1,-1/\tau)$ and. $(-1/\tau,-1/\tau^2,1)$.

In all cases this indexation was found to correspond with identified scattering planes in the structures found in a model like figure 1. As before, the indexation is extended radially by multiplying by τ^m and interstices in the pattern can be completely indexed by the usual vector addition. The atomic map of a first order supercluster is shown in figure 5b. To make up the indexation for axes through the centers of 20 faces, there is an additional equivalent for this 3-fold axis, namely $[1\tau\tau]\sqrt{3/(1+2\tau^2)}$, where the indices take positive or negative values and the square root normalizes to the same radius as [111]. In the following 'structure factor' analysis, icosahedral symmetry takes care of the additional numbers and only the simpler form is needed.

We index in 3 dimensions to describe the geometric orders because our basic system and law is axial. An illustration of the geometric series is applied to 3-dimensional indexation as follows. In the quasicrystal, the axial indexation on the 5-fold pattern in the direction towards a 2-fold axis progresses:

$$(0,0,0)..(2/\tau^2,0,0),(2/\tau,0,0),(2,0,0),(2\tau,0,0),(2\tau^2,0,0).$$

(Off-axis, see the indexation of axial patterns in following tables, using conventions common in electron microscopy.) Compare with Bragg orders in the fcc Al matrix:

$$(0,0,0),(2,0,0),(4,0,0),(6,0,0),(8,0,0).$$

In the former case, the 3 dimensional indexation is descriptive and simpler than the alternative hyperspace description. The simplification was instrumental in calculating all of the structure factors for the beams occurring around the three major axes in the pattern. The indexation is therefore complete.

The projections that we make correspond to reciprocal lattice construction in crystalline specimens [15]. Whereas, the 5-fold and 3-fold patterns, can be completely indexed by vector extensions of the "third bright ring", the 2-fold pattern extension is made from the two crosses to be described.

This pattern (Schoenflies D_2) is therefore more complicated, and early failed attempts at simulation recorded special difficulty [18]. Firstly, the pattern is dense, suggesting an occurrence of double diffraction. The map in figure 5c shows

Figure 5a. 5-fold view from $[0\bar{1}\tau]$, with aligned planes of Mn (filled circles) and Al (unfilled), in a supercluster order 1 .

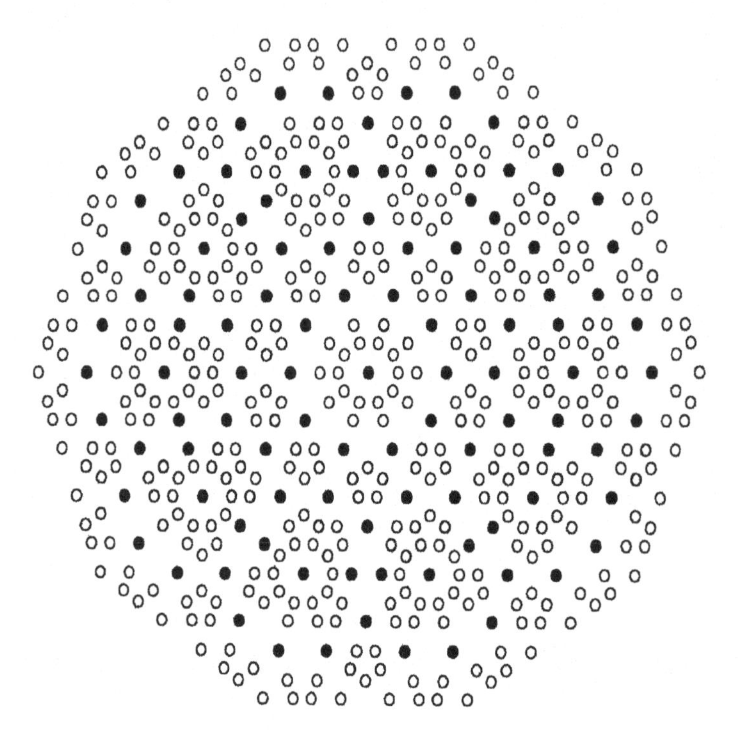

$[01\tau]$

special features that relate to coherence of scattering planes. The lines that are shown joining atoms, map subcluster bonds. The vertical and horizontal planes align in ways that are typicalin the 5-fold and 3-fold maps, but the diagonals are different. It seems that misalignment results in incoherence between planes separated in geometric series, allowing diffraction to occur only in arithmetic series. Beyond the charted observation, this difference is simulated. The pattern therefore contains two parts, one in geometric series; the other in arithmetic series. In order to understand how the two patterns combine, it is necessary to notice one further feature of quasicrystal diffraction that is evident especially in foils that are thin (we return to this in section 7). In electron diffraction generally, the deviation parameter on the Ewald sphere construction causes near axial reflections to be bright, the brightness falling with increasing deviation parameter until the occurrence of the first order Laue zone. By contrast in quasicrystals, the brightness of diffraction patterns typically increase axially to about the third 'bright ring', and then fall away without excitation of higher order zones [6, 19] (part of [6] is copied in section 7 below).

The foregoing reading of the map correlates with an inspection of the 2-fold diffraction pattern [2]. In figure 6 the cross is shown to repeat on the third bright ring on the diagonal X (right), while the cross-arms repeat on the mid-row left. Following indexation of each diffraction point on both the cross and diagonal X, complete indexation of the otherwise complex pattern turns out to be straight forward and will be listed in a later section. As with the 3-fold pattern, there are additional indexations for the axes that correspond to midpoints on the 30 sides. The 24 additions are accessible by unitary transformations. For the following 'structure factor' calculations, the simplest equivalent is sufficient.

Figure 5b. 3-fold [111] view of a supercluster order 1, with aligned planes.

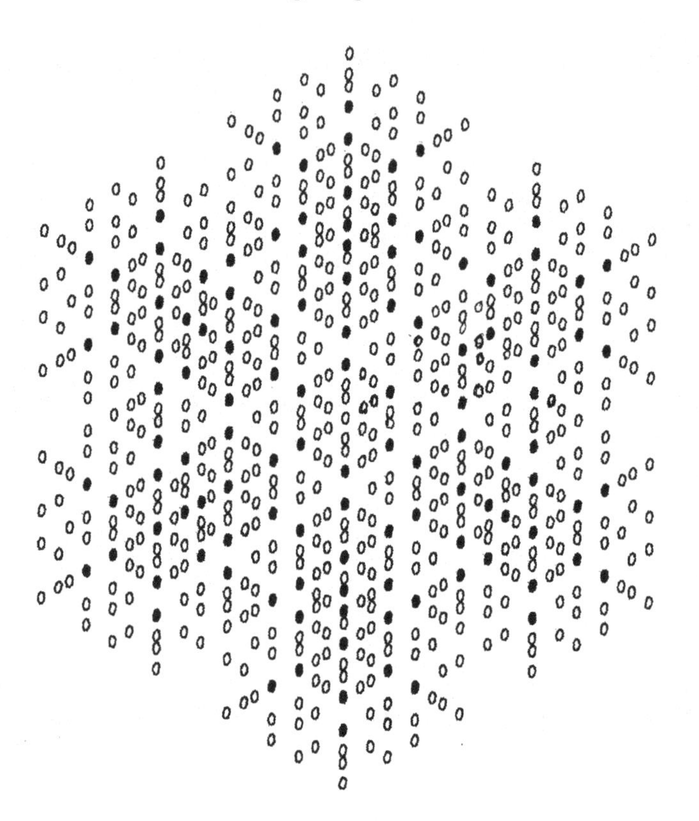

[111]

Figure 5c View of supercluster order 1 from the 2-fold [001] axis. Lines show subcluster bonds. The intercellular geometric series diagonal planes are misaligned; vertical and horizontal geometric series planes, as also the intracellular linear diagonal planes (with spacing d_d) are aligned. The misaligned planes do not diffract.

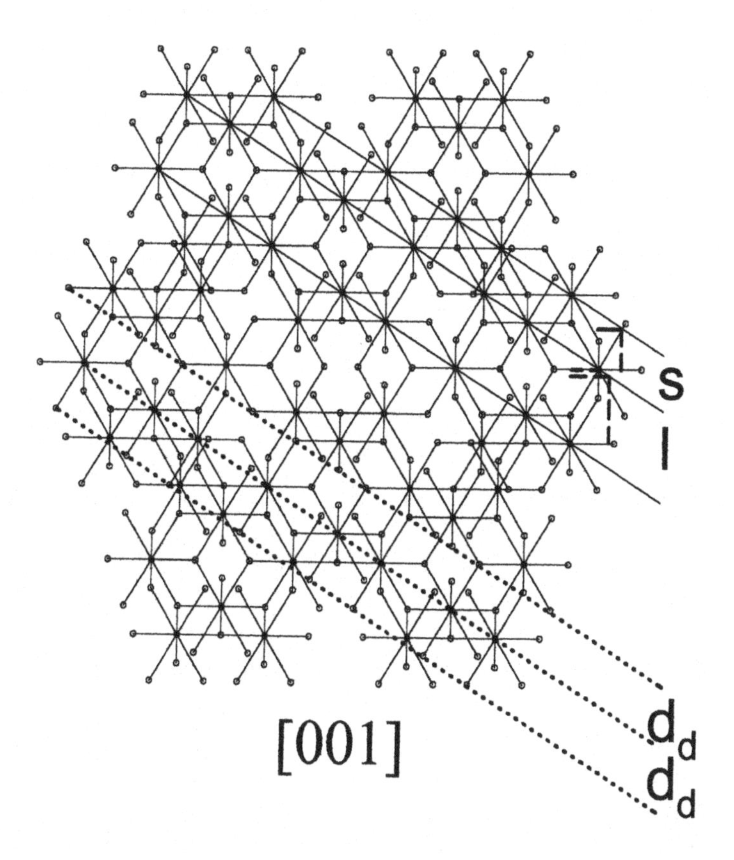

→

Figure 6. Composition and indexation of 2-fold [100] axial diffraction pattern (bottom) for modeled i-Al_6Mn, based on a geometric cross, and Bragg diagonals (top) with linear orders. The cross arms superpose (center row left) on the verticals and horizontals, and (center right) on the diagonals and their 3rd bright spots (see text). The two systems 'double diffract' (center) and superpose to match Shechtman's data [2,][xii]. Indexation follows straightforwardly. The method is three dimensional.

Electron microscopists normally use three dimensions, as in their indexation of the hexagonal close packed structure [15, appendix 5], where X-ray uses four [20]. Indexation requirements are different in electron microscopy from requirements in X-ray diffraction. Correspondingly, methods used are different. Our three dimensional indexation is based, for ease of application, on a subgroup (T_d), of the icosahedral point group symmetry (in the hierarchic model).

[xii] It is easier to observe the patterns than to describe them, so look again if you don't get it first time. The point is that the complex pattern at the bottom is derived by repetition of the simple patterns at the top, both easily indexed.

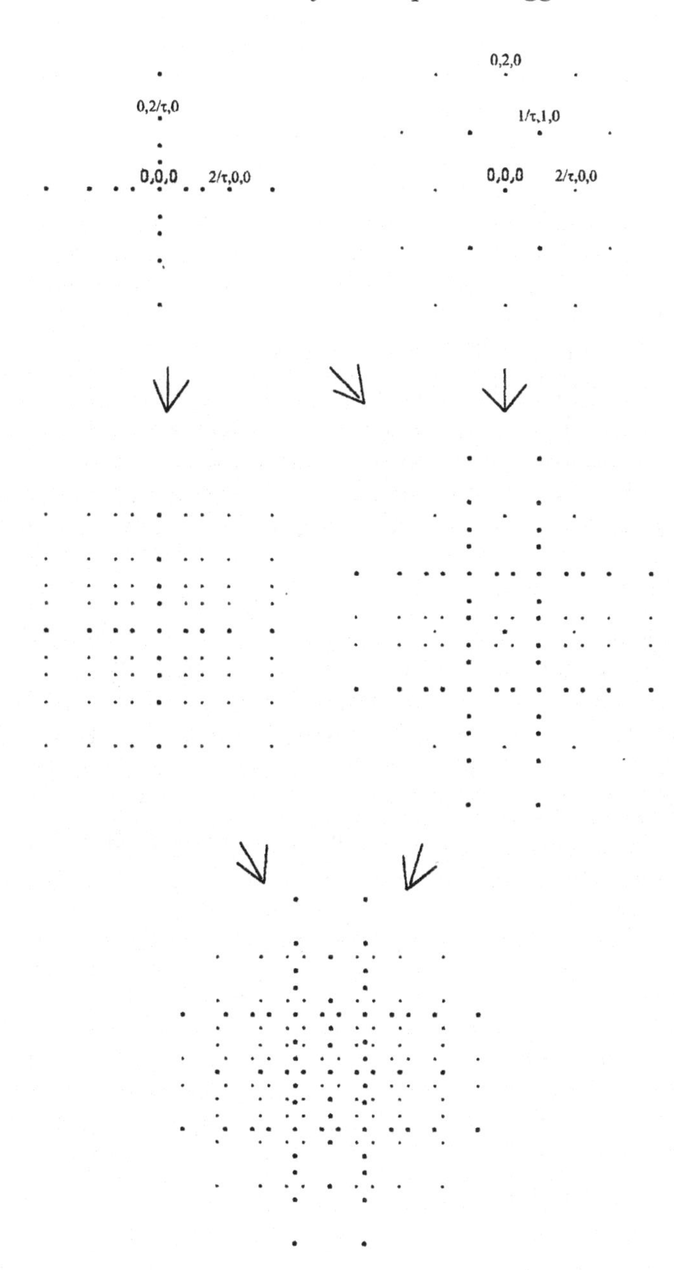
0,2/τ,0
0,0,0
2/τ,0,0
0,2,0
1/τ,1,0
0,0,0
2/τ,0,0

With this basis of atomic maps and planes, we need next to ask what is required to calculate scattering intensities for either electrons or x-rays. Bragg planes are evident in figure 5. We consider first similarities with crystals and then differences. On crystal planes the Bragg diffraction is specular including the plane normal which coincides with a vector that joins two beams in reciprocal space. This vector is used to label the corresponding reflection. The scattering amplitude is the sum of the amplitudes of all rays scattered by the crystal atoms. Since a crystal is periodic the summed amplitudes due to the individual cells are in phase (for a given index of reflection) and the scattering then depends on the structure factors of the unit cells. The incident and diffracted beams are sinusoidal waves. How do these waves interact with a geometric solid?

In the supercluster, decoration of the icosahedral cells is defined and so is the regularity in the placement of the cells. It is therefore feasible to treat the LPS as a large quasi unit cell in order to calculate its 'structure factor' using conventional formulae. Except for the quasi reciprocal lattice vectors provided by indexation, reciprocal quasi-lattice cells are undefined (they approximate to quasi-Brillouin cells in logarithmic space) and space in the geometric lattice is not necessarily filled by them.

To illustrate the method we first calculate the structure factor for the $2/\tau,0,0$ reflection for a supercluster order 6. This reflection occurs in both the 5-fold pattern and in the 2-fold pattern. Initially, and to aid the computation by reducing truncation effects after a large number of additions, we introduce form factors to represent individual orders of supercluster. (However, in later computations we shall revert to the simpler formulation and reduce the cluster size to order 2.) Meanwhile, following the crystalline case, a 'structure factor' F_{hkl} for a reflection with quasi Miller indices h,k,l, can be written [20]:

$$F_{hkl} = \sum_j f_j^{\varsigma} \cos(2\pi(hu_j + kv_j + lw_j)) \tag{2}$$

summed over atoms *j,* each with atomic scattering amplitude f_j^{ς}, on coordinates u_j, v_j and w_j, where the superscript ς identifies the chemical species on the site: Al or Mn. There are thirteen atoms in the subcluster, arranged at the center and at the corners of a regular icosahedron. The arrangement is centro-symmetric. The equation is simple to apply because our axes are Cartesian. In the model for quasicrystalline i-Al_6Mn, atomic sites in a subcluster are located at the 12 permutations of $(0,\pm 1/2,\pm\tau/2)$ for Al and (0,0,0) for Mn. About a cluster center, subcluster centers are located at the 12 permutations of $(0,\pm\tau/2,\pm\tau^2/2)$, and these values are multiplied by τ^{2p} for equivalent locations relative to the center of a supercluster order *p*. The sites are hierarchically ordered. With these numbers all atomic sites are located. Notice that equation (2) represents a Fourier transform.

Using equation (2) we begin by calculating scattering amplitudes for subclusters and the cluster and will later progress, through multiple steps, to high order superclusters. Sum over all 156 atomic sites using the same formula (2). Because of overlap and vacancies, the 156 sites contain only 104 atoms, 92 of them being shared Al sites on the outer shell or inner hole. The well-defined cluster, yields, summing equation (2) over these 156 sites, a cluster form factor $F_{hkl}^{cluster}$. The sites occur on three shells:

The central surface (where icosahedra share edges) is dodecahedral [5,8], containing 20 atoms on a radius of $\tau\sqrt{3}$. Each of these atoms lies on a triple point and is shared by three subclusters. They are counted three times, so the applied atomic scattering factor on each count is $f_{Al}/3$. Standard values were used [15].

The outer shell is pseudo space filling and each atom is either shared with neighboring clusters or is a 'hopping site'. Shared atoms are counted twice, each time with the default value of scattering amplitude f_{Al} /2. The same factor is applied to 'hopping sites' in clusters. They occur where two Al atoms on adjacent subclusters are separated by a distance less than the diameter of Al. Typical hopping sites have a nearest neighbor separation of $1/\tau$. The default has the advantage of improving convergence in scattering power values with increasing order of supercluster. The convergence is a consequence of reduced scattering power given to the numerous atoms on the extreme outer shell.

The inner 'hole' is also pseudo space filling and is underpopulated. This shell has 12 sites at the corners of a regular icosahedron, but the side length is less than the diameter of Al, so that there is allowed space for filling only three of the sites. The applied atomic scattering factor is therefore f_{Al} /4. (Images of the cluster centers are in consequence 3-fold symmetric [1,5,16].

For the many simulations listed in section 5, this procedure is enlarged for superclusters of order 2. To extend to order 6, we first use an intermediate step, and this for computational reasons. Taking the systematically calculated scattering factor for the cluster, we now progress to approximating for the superclusters. A supercluster order 1 is composed of 12 clusters in icosahedral configuration. Write for the reciprocal lattice vector $\underline{g}_{hkl} = 2\pi(h\underline{a} + k\underline{b} + l\underline{c})$, expressed on orthonormals, $\underline{a}, \underline{b}, \underline{c}$, corresponding to the diffracted beam with quasi Miller indices h,k,l. Let $\underline{r}_j$ be the vector pointing to the coordinates of atom j, so that we can incorporate these into the equation for a supercluster order 1 where the clusters are centered at $\underline{r}_j$:

$$F^1_{hkl} = \sum_{j1}^{12} \sum_{j}^{156} f_j^{\varsigma} \cos(2\pi((\underline{r_j} - \underline{r_{j1}}).\underline{g_{hkl}} + \underline{r_{j1}}.\underline{g_{hkl}})) \quad (3)$$

This is a kind of form factor for the supercluster order 1. Then, by expansion of the cosine addition and ignoring the phase information in the sine terms from the centro-symmetric solid, approximate 'structure factors' can be summed iteratively using, for order p, the formula:

$$F^{p}_{hkl} \approx \sum_{j1}^{12} F^{p-1}_{hkl} \cos(2\pi(\underline{r_{jp}} \cdot \underline{g_{hkl}})) \tag{4}$$

As one example from many results for the model, calculated scattered intensities from a supercluster order 6 are illustrated in figures 7a and 7b, on a linear scale firstly and secondly on a logarithmic scale. Two significant consequences follow. Firstly, the maximum (by orders of magnitude) shown in these two figures was obtained by simulating in second Bragg order, n=2 so that the function in equations 2 seq. is written $\cos(4\pi(\underline{r}.\underline{g}))$ etc. The reason for this will become apparent in the next section as will the reason for the second consequence: the peak is offset from the Bragg condition at step 1 in the graph. The offset, c_s=1/0.947, is found by scanning the value of the index used $(\tau/2,0,0) < h_i < 0.8(\tau/2,0,0)$. We call the offset the compromise spacing effect (CSE). The offset has implications for both the quasi Bragg angle and for the measured interplanar spacing: $d' = d_{Bragg}/c_s$. The denominator is a metric that relates pattern dimensions to structure. This effect is illustrated in figure 7 and in a schematic way elsewhere [17].

Notice that, in figure 7b, the main peak is three orders of magnitude more intense than any of the harmonics. The primary purpose of the form factor method was to overcome computational truncation limits, but subsidiary information is also obtained, as can be seen. The method is available because the model is well defined.

A second example is the Bragg type higher orders, such as $4/\tau,0,0$. In supercluster order 6, further simulations showed over two orders of magnitude intensity reduction compared to

Figure 7a. Simulated scattering power scanned away from Bragg angle for $(\tau/2,0,0) < h_i < 0.8(\tau/2,0,0)$ due to supercluster order 6, using the procedure described in text. Fractional divergence would be zero at nominal Bragg angle, and at the peak is c_s=1/0.947 Fuull scale is $6\text{x}10^{12}$.(Reprinted with permission of Nova Science Publishers)

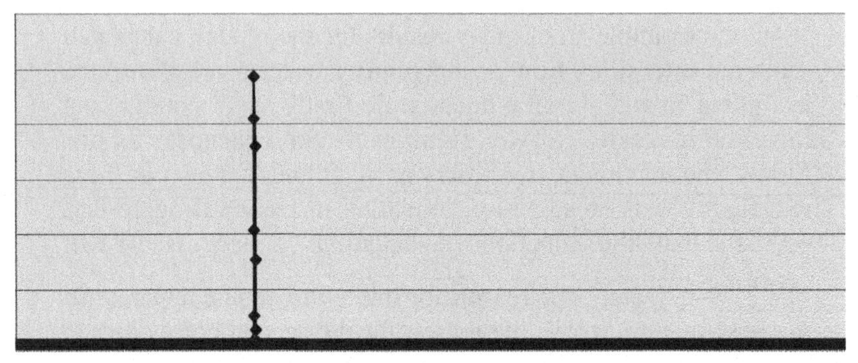

the fundamental $2/\tau,0,0$. The higher order is only marginally greater than other logarithmic harmonics shown in figure 7b. The result is therefore consistent with the general requirements for the theory, since higher (linear) Bragg orders are not experimentally observed, at least for this reflection. (By contrast, the diagonal beams in figure 6 are both exceptional and consistent with the quasi-Bragg law.)

Several tests were performed to check the consistency of the computations, and four are significant. Firstly, to confirm the cluster structure factor method of equation (4), diffraction simulations were compared using alternately equation (2) or equation (4). For supercluster order 3, the log profiles were the

Figure 7b. Logarithmic plot for the simulation of the $2/\tau,0,0$ diffracted beam as in figure 7a. The harmonics are identified: indices of the three strongest are, in order of intensity, $8/\tau^2,0,0$ at step 1911, $\tau^2,0,0$ at step 1057, and. $12/\tau^3,0,0$ at step 1383. (Reprinted with permission of Nova Science Publishers)

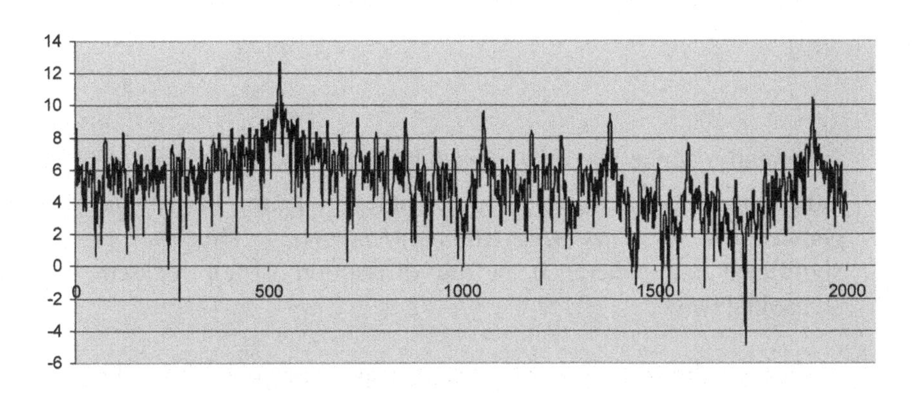

same except for a general increase in values in the latter case by less than an order of magnitude. Secondly, when profiles were compared for order 6 with profiles from supercluster order 3 using either of equations (4) or (2), the main difference was sharpening of the diffraction peaks and of the logarithmic harmonics. Examples will be given in the next section. Thirdly, a comparison of the $2/\tau,0,0$ profile with $1/\tau,0,0$ showed a decrease in computed values for the latter by four orders of magnitude. The fact is consistent with the following analysis of geometric series. Fourthly, when all of the Al atoms or all of the Mn atoms were ignored, the results shown in figure 7 were unchanged in form, except, naturally, for the overall simulated intensities. Finally, when the computed intensities to be listed below were compared, the results were consistent with rankings in experimental data [2] including the dense 2-fold pattern; the third bright ring; the 5-fold pattern and the 3-fold pattern..

3. Geometric and Fibonacci series

As described earlier, the geometric series, which more precisely describes diffraction patterns in quasicrystals, is a restricted Fibonacci series. Several consequences follow from the values shown in table I. It is clear firstly that the geometric series G_m is equal to the sum of two or more Fibonacci series F_m, e.g.:

$$\begin{aligned} G_m &= \tau^m \\ &= F_{m-1} + F_m \tau \\ &= F_{m-1} + F_m + F_m/\tau . \end{aligned} \tag{5}$$

Secondly, since $F_m \rightarrow F_{m-1}\tau$ as n increases, all terms are either integral or tend to integral values. *At long range, the geometrical series is approximately integral.* This fact is significant for coherence in the diffraction of an incident sinusoidal wave.

Thirdly, the fact is also a necessary component in the explanation of the CSE. Notice, in table I, the dominant effect of the spacings in the subcluster and cluster due to multiplicities, and also the divergence from zero of the values for mod(G_m,0.5). The simulation (figure 7) demonstrates that their combined and weighted effect in a supercluster of high order is ~5.6%. Thirdly, therefore, *the CSE is a local effect periodically repeated.* Table I represents the simple case for the $2/\tau$,0,0 reflection, but the result was extended to other indexed reflections.

This conclusion is reinforced by simulations of computed structure factors (summarized in table IIa and illustrated in figure 8) for several real and imaginary structures: a crystalline, face centered cubic solid; a logarithmically periodic solid; and two intermediate hypothetical variants one an icoasahedral cell on an fcc lattice; the other *vice versa.* In each case the number of atoms in the sample size was equivalent to a supercluster order 3. The asymmetry in simulated CSEs shows that logarithmic spacings trump linear spacings.

Table I. Geometric series containing two Fibonacci series tending to integral values at high index *n*. For the $2/\tau$,0,0 reflection, interplanar spacings in the subcluster from 0 $\pm$1 and $\pm\tau$. These are distributed by the spacings at cluster centers (0,$\pm\tau$,$\pm\tau^2$), and at supercluster order 1 (Sc1) by spacings (0,$\pm\tau^3$,$\pm\tau^4$) etc. Relative multiplicities are shown for a supercluster order 3. The geometric series tends to half integral values at small *n*, and integral values at large *n*.

Geo-metric G_n *	Fibonacci terms	Values	Mod(G_n,0.5)/f_n %	Containing structural spacings in:	-with relative Multi-plicities in sc3
τ^{-1}	$\tau-1$	0.618034	-	subcluster	20736
1	1	1	0	"	20736
τ	τ	1.618034	7.9	" & cluster	22464
τ^2	$1+\tau$	2.618034	4.5	cluster	1728
τ^3	$1+2\tau$	4.236068	5.6	Supercluster Order1	144
τ^4	$2+3\tau$	6.854102	-2.1	Sc1	144
τ^5	$3+5\tau$	11.09017	0.81	Sc2	12
τ^6	$5+8\tau$	17.94427	-0.31	Sc2	12
τ^7	$8+13\tau$	29.03444	0.12	Sc3	1
τ^8	$13+21\tau$	46.97872	-0.04	Sc3	1
τ^9	$21+34\tau$	76.01316	0.02	Sc4	-
τ^{10}	$34+55\tau$	122.9919	0.006	Sc4	-
τ^{11}	etc.	199.005025			
τ^{12}		321.9968944			
τ^{13}		521.0019194			
τ^{14}		842.9988138			
τ^{15}		1364.000733			
τ^{16}		2206.999547			
τ^{17}		3571.0002			
τ^{18}		5777.999827			
τ^{19}		9349.000107			
τ^{20}		15126.99993			
τ^{21}		24476.00004			
τ^{22}		39602.99997			
τ^{23}		64079.00002			

* actual spacings are ½, $\tau/2$ $\tau^2/2$ … etc.

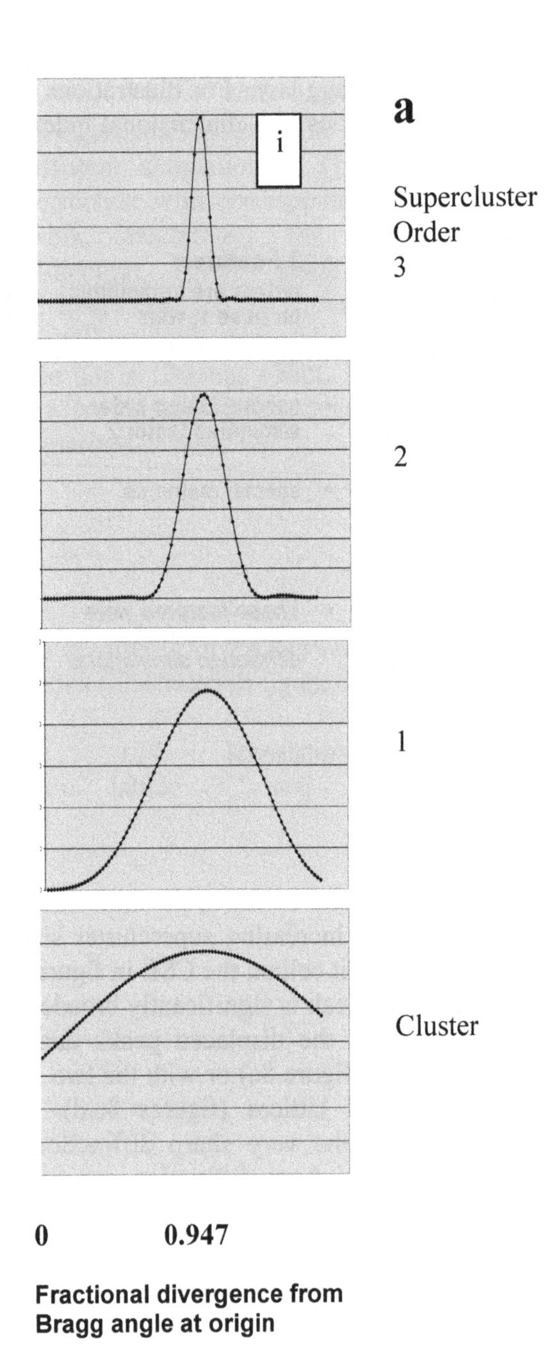
i
a
Supercluster
Order
3
2
1
Cluster
0
0.947
Fractional divergence from
Bragg angle at origin

←

Figure 8 (a). Computed intensities, i.e. squared structure factors, for $(2/\tau, 0,0)$ diffraction in supercluster orders 3,2,1, and in an i-Al_6Mn cluster. The intensities are plotted against divergence from Bragg angle and span ~10% of this angle. The Compromise Spacing Effect (CSE) shifts the calculated Bragg angle by 5.6%. The lines from different supercluster orders show the same CSE displacement from the Bragg angle, and they narrow with increasing order.

→

Overleaf **figure 8 (b)** (111) structure factors, computed against angle as above, for a cubic cluster of fcc Al, having about 19,000 atom sites, similar to quasicrystal supercluster order 2. There is no CSE. At this size, the peaks are computed to be comparatively broad, typically 5% of the Bragg angle. **(c)** Computed structure factor for (111) diffraction by an imaginary cubic fcc cell in a logarithmically periodic icosahedral grid, as in supercluster order 2. The CSE occurs with a peak narrower than in figure 8b. **(d)** Computed structure factor for (τ, τ, τ) i-Al_6Mn icosahedral cell in an imaginary cubic grid of side τ and site population about 20,000 (like b and c)

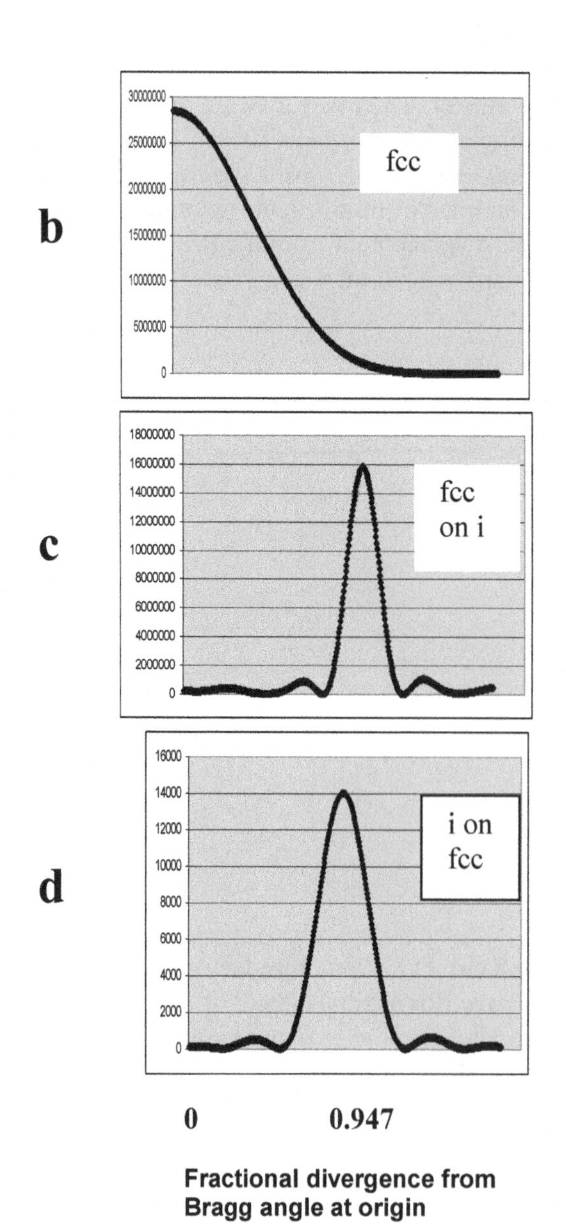
b
fcc
30000000
25000000
20000000
15000000
10000000
5000000
0
c
fcc
on i
18000000
16000000
14000000
12000000
10000000
8000000
6000000
4000000
2000000
0
d
i on
fcc
16000
14000
12000
10000
8000
6000
4000
2000
0
0
0.947
Fractional divergence from
Bragg angle at origin

4. Detailed indexation and measurement

With a detailed model in which every atom is specified and located, we now consider how they are measured. As we have seen, the diffraction differs from Bragg diffraction because of the logarithmic orders and because of the half integral approximation at short range. Are measurements of interplanar spacings also affected? Consider the quasi-Bragg law

Start with the established law for crystals, and address the problem of measurement of lattice parameter for cells that do not fill space and where the reciprocal quasi-lattice is still to be defined. For convenience the unit of length is already taken as the edge length of the icosahedron, equal to the experimental diameter of Al. This has to be related to the third bright ring on the diffraction pattern from a thin foil (shown below in section 7). The measured interplanar spacing corresponding to *Bragg's law in first order* is d=0.206 nm [6,22]. Matching the preliminary indexation for the 5-fold pattern already described [5 appendix A, 22] the obvious correlation applies to an index of $h=2/\tau$, so that, for the $2/\tau$,0,0 reflection, the parameter $a=dh=0.255$ nm. This is short of the diameter of Al but is close enough (0.269 nm) when the CSE is applied, as in equation (6) below. So the modified Bragg law provides a measurement due to an unambiguous index. How does this relate to half integral coherence and the geometric sequence? Then the Bragg law operates in second order so that:

$$Bragg: \frac{2\lambda}{2\sin(\theta)} = 0.206nm = \frac{a}{\sqrt{\sum h_i^2 c_s}} = \frac{\lambda}{\sin(\theta)} : Quasi - Bragg \tag{6}$$

where a is the (cubic) cell size. For the reflection $2/\tau$,0,0, the icosahedral edge is measured to be 0.269 nm, as before. Notice that wavelength is related to interplanar spacing through the quasi Bragg law, whereas the structure factor depends on direction cosines.

If the icosahedral edge is slightly shorter than the typical diameter of metallic Al, notice that the solvent atoms are dense and, by presumption, donate charge to the central solute.

Moreover, the unit icosahedral cell is presumed bulbous because it shares edges, so the effective diameter of Al is again reduced. The volume of the icosahedral cell is $5\tau^2/6$. The real space volume of the subcluster is 17% smaller than the volume occupied by 13 equivalent atoms in the face centered cubic matrix (figure 4a). The high density of the icosahedral subcluster is the obvious driving force for the structure. Meanwhile, as we shall see in simulations like those to be described in the following section, there is an excellent fit when the structure factors are calculated from $F_{hkl} = \sum_i 4\pi i\,(\mathbf{h}_i.\boldsymbol{r}_i)$ summed over atoms in three dimensions, subscript i=1,2,3.

5. Beam intensity rankings.

Many circumstances affect intensities measured in electron or X-ray diffraction. In electron diffraction, specimen orientation, and consequent extinction distance are, in particular circumstances, critical. Multiple scattering, temperature, specimen purity, beam uniformity, lens aberrations and other factors also influence measurements. If a specimen is sufficiently thin, the diffraction may be studied in the kinematic rather than dynamic approximation, and this is supposed in the following calculations.

Table III and figure 9a illustrate the decagonal pattern, indexed and analyzed around the 5-fold axis. The second order supercluster that is simulated is larger than the first order map in figure 5a. The pattern shown in figure 9a is radially correct but omits intermediate reflections between spokes. These can be easily inserted by comparison with published data, indexed by vector additions and their intensities simulated. Similarly, in table IV and figure 9b, the diffraction about the 3-fold axis is indexed and analyzed. The pattern may be compared, for angular symmetry, with the map in figure 5b. The pattern from the 2-fold axis is more complex and is dense in features. Analysis presents two particular problems. The first is: with which of the two inconsistent patterns published by Shechtman et al., [2] should simulations be compared? The second is how to index them?

Figure 9a. Indexation of the 5-fold axial diffraction pattern in quasicrystalline i-Al_6Mn with symbols listed in the following table I. Corresponding structure factors are given in the table.

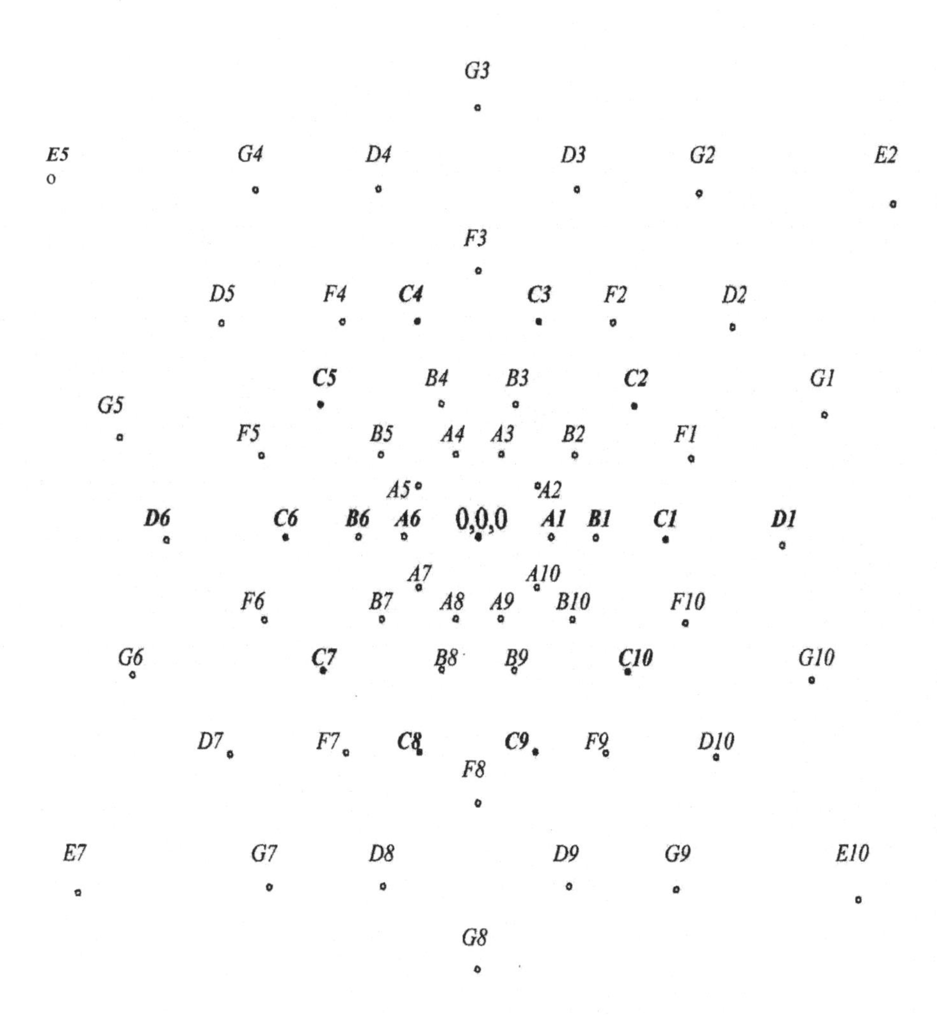

Table III. Indices for 5-fold axial pattern and squared structure factors, corresponding to figure 9a

Symbolic index	h	k	l	Intensity*	Comment
A1**	$2/\tau^3$	0	0	20k	$C1/\tau^2$
B1	$2/\tau^2$	0	0	300	$C1/\tau$
C1	$2/\tau$	0	0	570	
C2	1	$1/\tau$	$1/\tau^2$	"	
C3	$1/\tau^2$	1	$1/\tau$	"	
C4	$-1/\tau^2$	1	$1/\tau$	"	
C5	-1	$1/\tau$	$1/\tau^2$	"	
C6	$-2/\tau$	0	0	"	
C7	-1	$-1/\tau$	$-1/\tau^2$	"	
C8	$-1/\tau^2$	-1	$-1/\tau$	"	
C9	$1/\tau^2$	-1	$-1/\tau$	"	
C10	1	$-1/\tau$	$-1/\tau^2$	"	
D1	2	0	0	550	$C1.\tau$
E1	2τ	0	0	390	$C1.\tau^2$
F1	$3-\tau$	$2-\tau$	$2\tau-3$	76	C1+A3
F2	$3\tau-4$	1	$\tau-1$	"	C3+A1
F3	0	$2/\tau$	$2/\tau^2$	"	C2-D1+C2
F4	$4-3\tau$	τ	$\tau-1$	"	A6+C4
F5	$\tau-3$	$2-\tau$	$2\tau-3$	"	C6+A4
F6	$\tau-3$	$\tau-2$	$3-2\tau$	"	
F7	$4-3\tau$	-1	$1-\tau$	"	etc.
F8	0	$-2/\tau$	$-2/\tau^2$	"	
F9	$3\tau-4$	-1	$1-\tau$	"	
F10	$5-\tau$	$\tau-2$	$3-2\tau$	"	
G1***	0	2	$2/\tau$	230	$F1*\tau$

* Structure factor squared: peak height x FWHM, for a supercluster order 2, in units of 1000.
** A2, A3, A4… similar to $C2/\tau^2$, $C3/\tau^2$, $C4/\tau^2$ … etc
***G2, G3, G4…similar to $F2\tau$, $F3\tau$ $F4\tau$ …etc

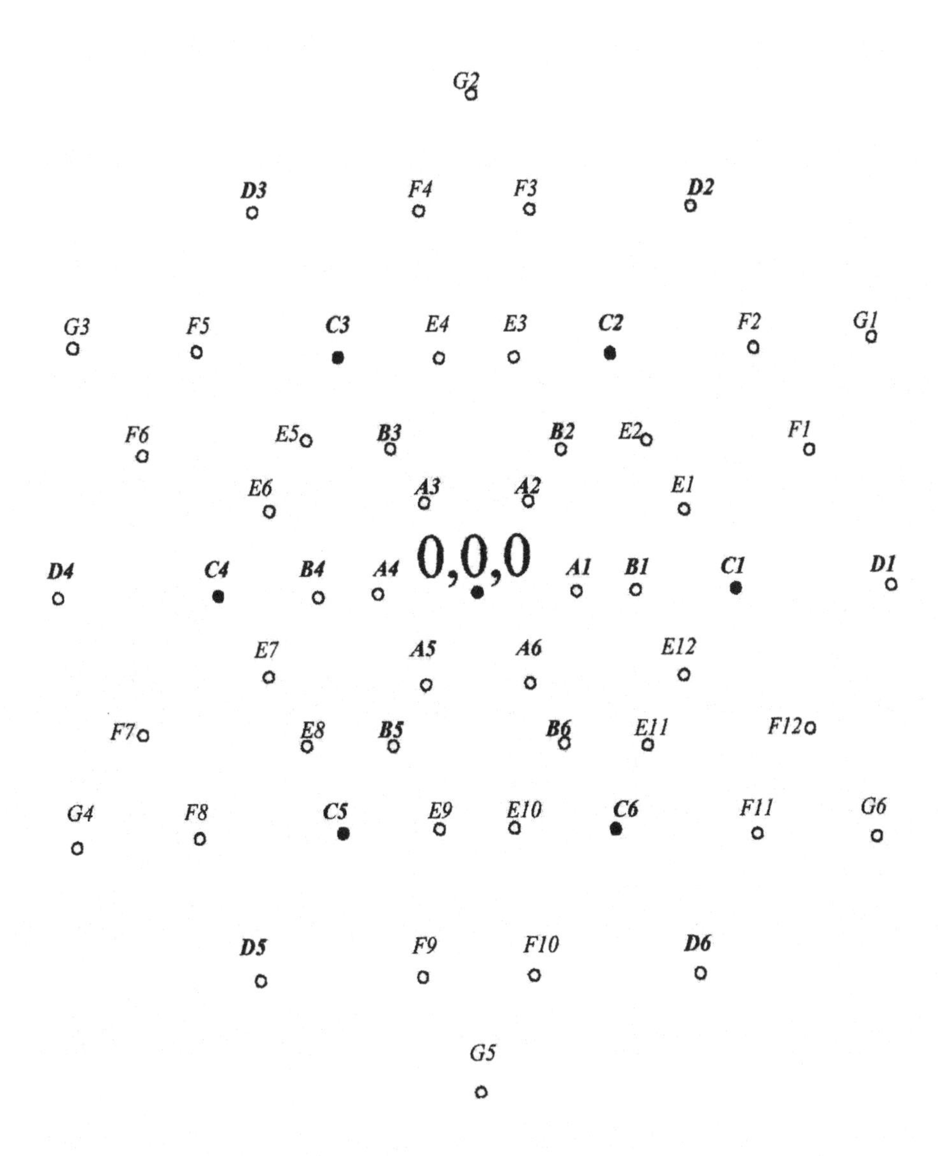

Figure 9b. Indexation of the 3-fold axial diffraction pattern in i-Al_6Mn and squared structure factors computed for a supercluster order 2. (see table IV)

Table IV. Computed intensities for 3-fold axial pattern, corresponding to figure 9b. (* Structure factor squared: peak height x FWHM for supercluster order 2 (in thousands)

Symbolic index	**h**	**k**	**l**	**Intensity***	**Comment**
A1**	$-1/\tau^2$	$1/\tau^3$	$1/\tau^4$	20k	$C1/\tau^2$
B1	$-1/\tau$	$1/\tau^2$	$1/\tau^3$	300	$C1/\tau$
C1	-1	$1/\tau$	$1/\tau^2$	580	
C2	$-1/\tau^2$	1	$-1/\tau$	"	
C3	$1/\tau$	$1/\tau^2$	-1	"	
C4	1	$-1/\tau$	$-1/\tau^3$	"	
C5	$1/\tau^2$	-1	$1/\tau$	"	
C6	$-1/\tau$	$-1/\tau^2$	1	"	
D1	$-\tau$	1	$1/\tau$	546	$C1.\tau$
E1				26	B1+A2
E2				"	A1+B2
E3				"	A3+b2
E4				"	B3+A2
E5				"	C4+B2
E6				"	C4+A2
E7				"	C4+A6
E8				"	C4+B6
E9				"	C5+A1
E10				"	C5+B1
E11				"	C6+A2
E12				"	C6+B2
F1				200	C1+B2
F2				"	B1+C2
F3				"	B3+C2
F4				"	C3+B2
F5				"	B4+C3
F6				"	C4+B3
F7				"	C4+B5
F8				"	B4+C5
F9				"	C5+B6
F10				"	C6+B5
F11				"	C6+B1
F12				"	C1+B6
G1	$\tau-3$	τ	$-1/\tau^2$	79	C1+C2
G2	$1/\tau^3$	$3-\tau$	$-\tau$	"	
G3	τ	$-1/\tau^3$	$\tau-3$	"	
G4	$3-\tau$	$-\tau$	$1/\tau^3$	"	
G5	$-1/\tau^3$	$\tau-3$	τ	"	
G6	$-\tau$	$1/\tau^3$	$3-\tau$	"	

Figure 9c Indexation convention derived from the double diffraction illustrated in figure 6 and used in structure factor calculations for the 2-fold pattern in tableV. Brighter diffracted beams are filled. They include the cross and the Bragg diagonals.

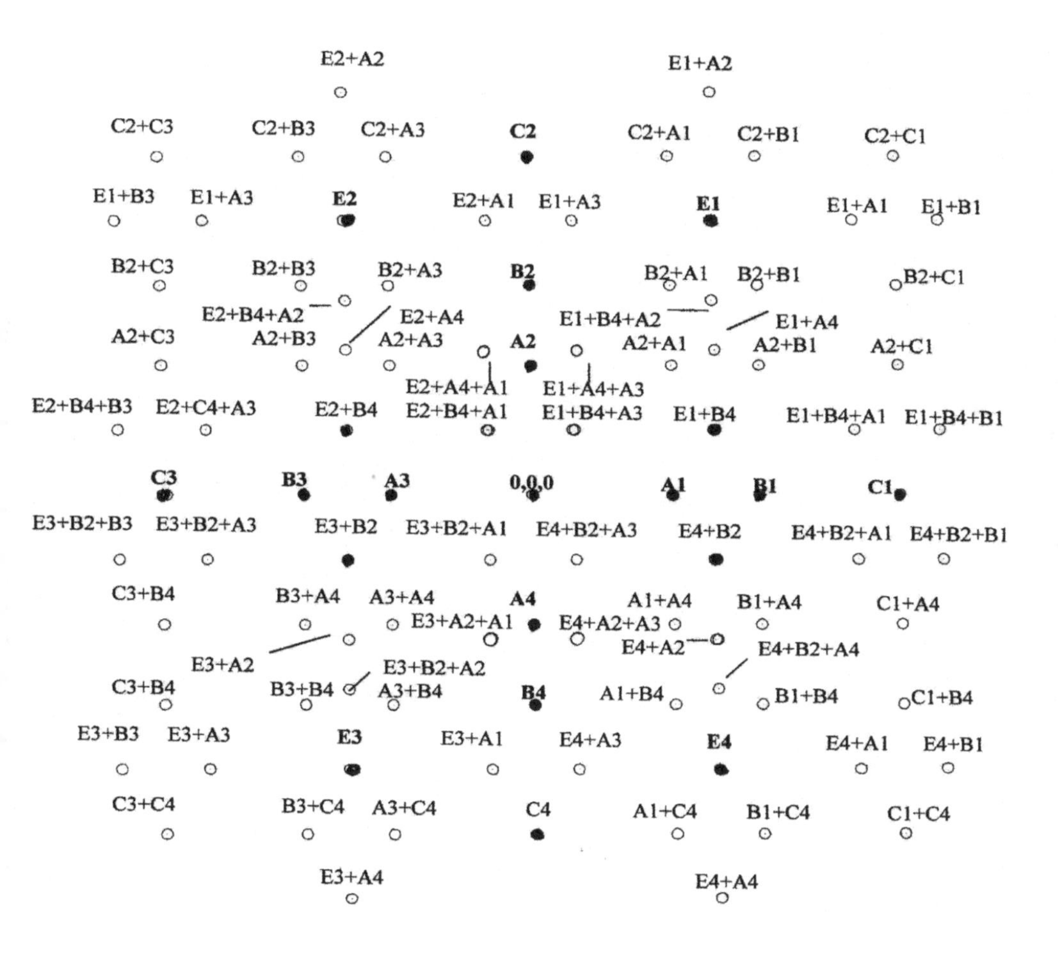

Table V. Excellent match between calculated intensities with experimental values [2] ranked for the beams indexed on the positive quadrant in figure 9c

Symbolic Index	**h**	**k**	**l**	**Intensity***	**Experimental rank**
	0	0	0		10
A1	$2/\tau^3$	0	0	20k	6
A2	0	2τ	0	"	6
A3	$-2/\tau^3$	0	0	"	6
A4	0	$-2/\tau^3$	0	"	6
B1	$2/\tau^2$	0	0	300k	7
C1	$2/\tau$	0	0	570k	9
D1	2	0	0	550k	8
E1	$1/\tau$	1	0	960k	9
F1	$2/\tau$	2	0	234k	5
F5	$3/\tau$	1	0	132k	5
E1+B4+B1				89	3
E1+B4+A1				57	3
E1+B4				545k	8
E1+B4+E3				37	Close to 0-order
A2+C1				4.3	4
A2+B1				2.6	4
A2+A1				230	3
E1+A4				230	3
E1+A4+A3				1.4k	3
E1+B4+A2				680	1
B2+C1				75k	5
B2+B1				25k	4
B2+A1				250	3
E1+B1				330k	5
E1+A1				1.2k	3
E1+A3				780	5
C2+C1				200k	6
C2+B1				70k	4
C2+A1				1.9k	2
E1+A1				15k	5
E1 Rotated 90°	Forbidden			17	?

* Structure factor squared: peak height x FWHM, for a supercluster order 2.

The first question will be answered via the second. It is evident, after inspection, that the 2-fold pattern contains two components: a vertical-horizontal cross with logarithmic periodicity; and a diagonal cross with linear, or Bragg, periodicity. Figure 6 showed how the cross repeats by superposition of itself on the third bright ring and repeats itself again on the diagonals. The two patterns superpose. We have previously discussed this superposition in terms of evidence of double diffraction [8 ch.4] but this mechanism is not important for the present purpose. The observation of superposition simplifies indexation: the two crosses are easily indexed individually, and then all the reflections are indexed by vector additions in the normal way. Such a result is shown in figure 9c and table V. There is an excellent match between the computed intensities (squared structure factors) and the experimental data [2] when reflections are ranked for intensity.

The bottom row of table V gives the key to the inconsistency in the originally published 2-fold patterns. The symmetry of the icosahedron requires that the 2-fold pattern located immediately between 3-fold axes, i.e. on the 5-3-2-3-5 axis, is normal to the 2-fold pattern on 5-2-5. The row, which is the simulated intensity for the first diagonal reflection on 5-2-5, shows intensity below noise. Assuming the model sufficient, then this 5-2-5 pattern has been wrongly transcribed. It should rotate by 90 degrees for the data to become truly icosahedral. The 2-fold pattern on 5-3-2-3-5 is correct. (Further illustration is given in [5], appendix D). The prediction is verifiable.

6. Modeled micrographs

The simulations so far discussed have been performed on the ideal hierarchic structure. Now we turn to real images of foils. We will model them in real space and, to check consistency, simulate their diffraction in momentum space. The model in figure 10b matches the easily identified clusters in figure 10a. In this image recorded with quasicrystal oriented on to the 5-fold axis, a cluster is identified by 10 subclusters arranged on a circle surrounding a tri-partite structure at its center. The measured diameters of the clusters and supercluster match the cluster model in figure 1.

Not only do the dimensions on the micrograph match the model, but simulated diffraction patterns of the model show systematic diffraction [8]. Meanwhile, attempts to model the tripartite center are reasonably successful [17 ch.6]. Even so it is not clear from the micrograph whether the regularity extends to the second order of supercluster. One implication is that disorder is common at this level of structure. We discuss further defects in the following section. Overall, the diffraction calculated for the model is similar to that calculated for a hierarchic model with a similar number of atoms. To obtain this kind of image, it is necessary to observe a sufficiently thin foil under optimum defocus conditions.

Figure 10 (a). TEM of i-Al_6Mn [1] recorded at optimum defocus. 3-fold centers [5] of regular decagons mark the cluster centers. Image reproduced with permissions from Bursill, Peng and from Nature, permission #2047130461239. **(b)** In mirror image of (a), skeletal structures representing icosahedral clusters, match the TEM image. Each corner of the golden rectangular triads locates an icosahedral subcluster containing central Mn + 12 *Al* atoms. Note the corresponding sections of supercluster order 1 outlined in each image with a pentagon that connects cluster centers.

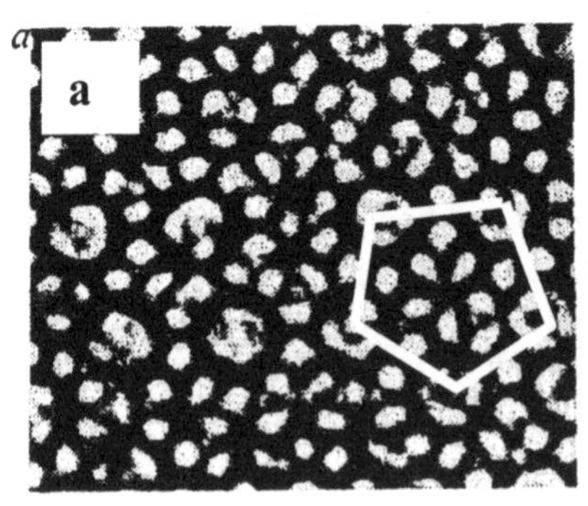

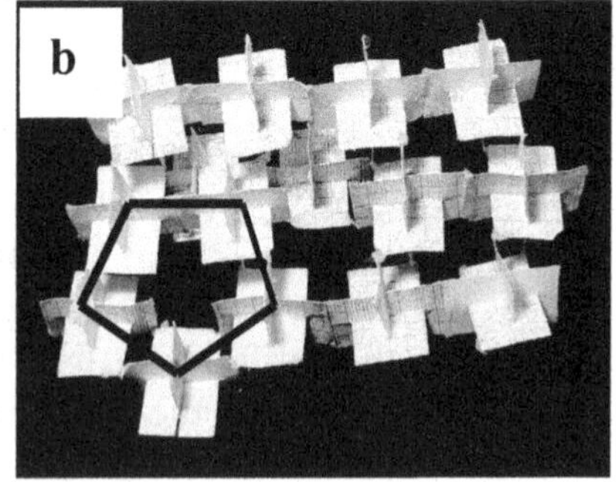

Notice that high resolution electron microscope imaging requires diffraction beam selection and image reconstruction [15]. Simulation is a necessary part of unambiguous structural determination Because of the column approximation and phase problem, neither methods nor programs are available for simulating quasicrystals.

Before proceeding to discuss defects in the structure, it is useful to summarize the novelty of the description by comparison with the methods of classical crystallography and with quasicrystallography (table VI). Most of these differences are mentioned or implied in the foregoing text.

7. Defects and 'holes'

Whether the LPS is an approximant or whether the quasicrystal is defective are less significant for the present model than the information given for the interaction of electromagnetic waves with atoms located in geometric series. This paper is written principally to illustrate how, in almost conventional terms, the diffraction occurs. Its origin was an attempt to understand a planar defect [6]. At that time the referee proposed delamination as the cause of interference fringes observed in the 5-fold convergent beam electron diffraction pattern. However several factors subsequently came to light suggesting that explanation was improvised. The planar quad is an obvious route to planar defects and to other disorder. Another type of defect was pointed to by Pauling [10]: icosahedra, while found within the unit cells of known crystal structures, have 'holes', though we find these only become problematic in the LPS at higher orders of supercluster. A proposed solution is due to overlaps for hole filling in superclusters [5]. Additionally, regular overlaps can explain the structure of decagonal quasicrystals to be discussed in section 8.2. The regular hierarchical LPS is most valuable for understanding mechanisms in quasi Bragg diffraction, but it is also useful for interpreting less regular structures that might seem to appear in the thin sections recorded in electron micrographs. To preserve the conditions for sharp quasicrystal

Table VI. Summary of comparisons of three conventions

Physical property	Hierarchic model	classical crystallography	quasi-crystallography
Bragg's law	adapted	yes	nominal
No. unit cells	1 icosahedron	1	poly polyhedra
dimensions	3	3	6
new crystal definition	no	no	yes
driving force/enthalpy described	yes	yes	no
atomic size fit	yes	yes	not measured
geometric series	explained & simulated	n/a	inaccurate Fibonacci
law simulated	yes	yes	no
complete	yes	yes	redundant orders
cell contact	double edge sharing	Face or single edge in silica	face
phase problem	assumed yes, like silica	as in silica	unknown
metric	cs=0.947	1	arbitrary PAS!
metric uniform	yes	yes	no
metric explained	yes	yes	chosen
metric simulated	yes	yes	no
modulus simulated	yes	yes	no
long, short range moduli	yes	n/a	no
approximant	hierarchic	n/a	various crystals
structural defects	typically glassy	n/a	unknown (3)
Shechtman's pattern	known defective	n/a	ignorant of defects
simulate symmetry error	yes	n/a	no

diffraction patterns, it is reasonable to assume that subclusters must be aligned. This is achieved both by edge sharing and by overlaps.

An example of planar defect diffraction is given in figure 11 and, schematically, in figure 12. Experimental details for the convergent beam electron diffraction have been described earlier [6], but the data is repeated here because reproduction available in modern publishing is greatly improved. The planar defect causes phase changes in diffracted beams that interfere above and below the defect. The modulation frequency in space varies proportionately with scattering angle. The phenomenon is similar to patterns observed, in transmission electron microscopy, from two films deposited one on top of the other [21]. This single demonstration is not exhaustive.

8. The hierarchic method

8.1 A hierarchic solution for icosahedral quasicrystals type II

The icosahedral phase i-Al_6Mn is typical for the binary quasicrystals. Firstly, its stoichiometry strongly suggests a small central atom surrounded by 12 larger atoms, shared with neighboring cells. The atoms match and fit. A 12-fold coordination around the small solute atom is optimum. The unit cell that we have previously described is internally consistent and leads naturally to the hierarchic model. The ratio of solute/solvent atomic sizes is fixed. Many icosahedral binary quasicrystals have similar stoichiometry and atomic size ratios. However, the original alloy is metastable, so other systems have been developed to discover more stable systems. A common route of exploration is to extend to ternaries, quaternaries, etc. Stability can be adjusted by these routes and single crystals grown, as we shall see. Often the stoichiometries and atomic size ratios have recognizable similarity to the original, but these systems are not examined here because of the added complication of composition and decoration. However some systems turn out especially different. One example is $Cd_{5.7}Yb$, significant because it can be grown apparently as large 'single crystals' suitable for X-ray diffraction studies [22].

Figure 11 (a) Convergent-beam electron diffraction (CBED) from 100 nm area of quasicrystal with contrast reduced at outer regions by photographic shadowing [6]. The central zero order, taken at quicker exposure is inset. **(b)** Corresponding selected area diffraction. Notice the 'third bright ring'. **(c)** same as (a) but with longer exposure to show structures of Kikuchi bands as arrowed, and **(d)** CBED with incident beam tilted off 5-fold axis. (wavy hairlines at right are printing defects.)

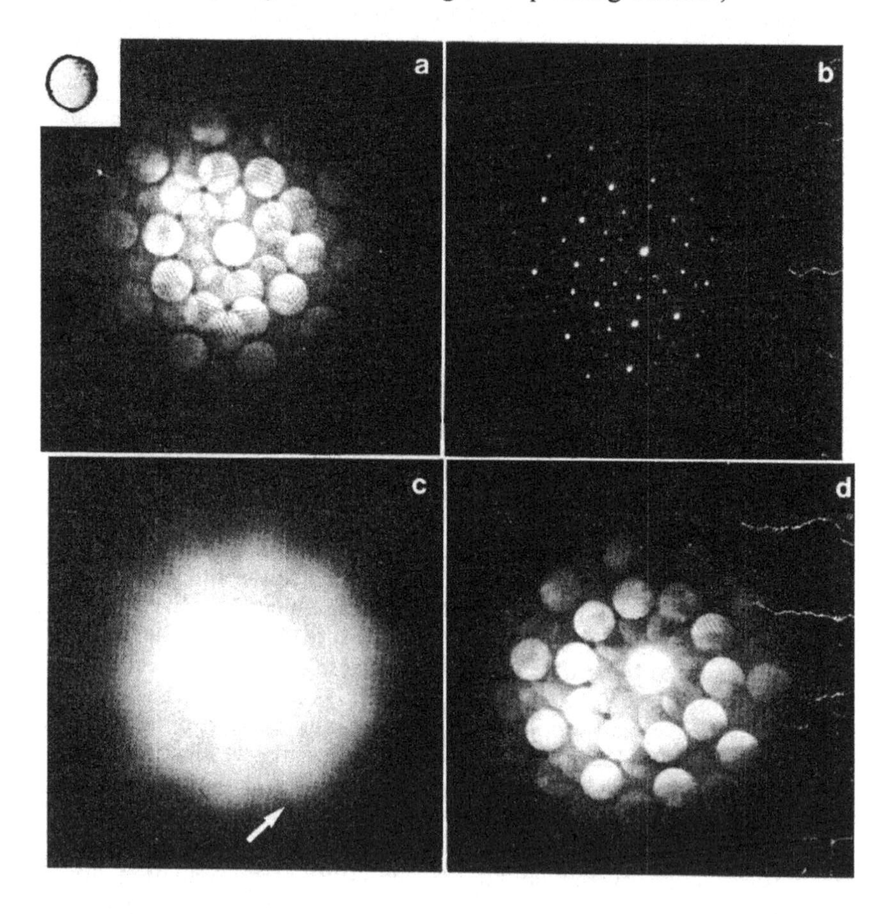

Figure 12. Drawing of interference fringes photographed in figure 11a. Generally, the fringes are tangential with fringe frequency inversely proportional to the scattering angle (approximately double the angle of quasi-Bragg diffraction). Some fringes appear to be double diffracted. The structure is typical for a planar defect normal to the mean incidence of the convergent electron beam.

Table VII. atomic sizes [23] in nm of metallic solvent and solute atoms in binary icosahedral quasicrystals of type I (Al_6Mn) and type II ($Cd_{5.8}Yb$). In type I the ideal ratio is 88%

Binary system	Metallic atomic radii of solute	solvent	Ratio
Al_6Mn	0.126	0.143	0.88
$Cd_{5.7}Yb$	0.192	0.154	1.25

Table VII shows that the atomic size ratios are incompatibly different from what has been described for i-Al_6Mn. Applying our icosahedral hierarchic method to understanding the structure of quasicrystals, it is first necessary to propose the unit cell for $Cd_{5.7}Yb$. This is surprisingly easy to do, using the same considerations that were used before. To match the diffraction pattern, search for a cell that is either icosahedral (preferably) or dodecahedral; that can be densely decorated; with atoms of matching sizes; and with correct stoichiometry. Because the ratio of solvent size to solute size is not only greater than the ideal 0.88 but considerably greater than 1, we have to find a radically new solution. In i-Al_6Mn, the stoichiometry provided a coordination of 12 around the solute atom. By contrast, in i-$Cd_{5.7}Yb$, because size requires the outer atoms to be shared with neighboring cells, close packing requires that the central atom must be a solvent atom and that the solute atoms are shared between cells, along with some of the solvent atoms. In order to preserve a dense structure it is clear that the outer sides must combine solvent and solute atoms. Now an icosahedron has 12 corners and 30 sides. For reasons of stoichiometry the next step is to suppose that the corners are occupied by solute atoms and the sides by solvent atoms. The unit side of the icosahedron is therefore given by adding respective radii: R_{Yb} +$2R_{Cd}$ +$R_{Yb.}$ From tables [24], this has a length about 0.692 nm, corresponding to the unit of length in our icosahedral units. Then it turns out that the diameter consisting of Yb-Cd-Cd-Cd-Yb, has a length that corresponds exactly with the diagonal of the golden rectangle. The corresponding structure is illustrated in figure 13.

The stoichiometry is calculated as follows: The thirteen atoms located internally are counted individually. The atoms on mid sides are shared and count for half. The atoms at the corners are shared, some at triple points and some only edge shared. A typical factor is between 2 and 3, say 2.5. The atom ratio is therefore given by Cd:Yb: ≈ (13 +30/2)/(12/2.5) = 5.8. The structure therefore matches all of symmetry, atomic size, close packing, and stoichiometry. As before the structure is understood to be hierarchic but defective. It includes both concave and planar quads. However the unit cells contain more atoms than in type I, *i.e.* 55 instead of the 13 that we previously discussed in i-Al_6Mn.

As shown in table VII, The ideal size ratio for type I is $R_{solute}/R_{solvent}$ $\sqrt{\tau^2+1}-1=0.90$. For type II the corresponding ideal ratio is $(3-\sqrt{\tau^2+1})/(1-\sqrt{\tau^2+1})=1.22$. Small differences from the values shown in the table are due to uncertainties in metallic size due to charge redistributions. Because of the consistency of the construction, this cell is a candidate for the unit cell structure for the type II i-quasicrystal. It is icosahedral, edge-sharing, (defective) hierarchical and avoids the multiplicities enjoyed by Takakura et al. [23] who use an alternative face-sharing construction. Since "Where are the atoms?" we now know "Why they are there?" No explanation is complete that fails to explain why the structure is not metallic glass. The firm rock of icosahedral structure (especially where it is confirmed by micrographs) dispenses with the leap of faith that suggests icosahedral diffraction from poly polyhedral cells (like those associated with rhombic triacontahedra [23]), which can be considered as defects. Accurate measurements are necessary to confirm the solution. Further work is needed to unravel regularities from defects, and it should be expected that electron microscopy with appropriate imaging, combined with X-crystallography from large samples, may continue to help. It is obvious that the type I systems are significantly different from type II. It may become possible to obtain better results by careful comparison and by employing the proper metric in dimensional measurement and interpretation.

Figure 13. A cut-out of a possible and consistent icosahedral unit cell for $Cd_{5.7}Yb$ that locates Yb atoms at the 12 corners of an icosahedral cell represented by golden triads. Three only are shown, represented by the larger darkened circles. Cd is located on the side centers, at the icosahedron center, and across the diagonals. The icosahedral side has a corresponding length about 0.692 nm[xiii] from typical metallic atom radii.

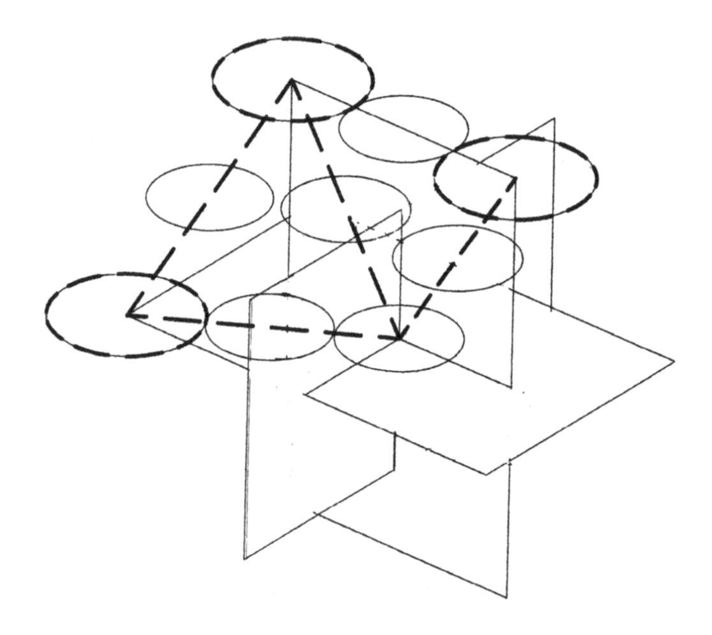

[xiii] If the lattice constant is, as reported, 0.5689nm, [25] and if the index is $2/\tau$, then the measured icosahedral side would be 0.7032 nm , with c_s=1. The unit cell would therefore be less dense than in type I, as is expected from the tangential rattle space in diagonal Cd.

8.2 Decagonal quasicrystals and others with reduced symmetry.

Decagonal quasicrystals display Bragg diffraction along one axis. Decagonal patterns occur on the plane normal to it. There are several ways in which a cluster can be sliced and shared. What matters for diffraction is that the structures align. Figure 14 illustrates one way in which this alignment may be possible. Two clusters share an apical subcluster

We find that this structure matches the dimensions in published micrographs [24], as do the imaged location of atomic mobility that correspond with hopping sites, *i.e.* sites on adjacent subclusters, only one of which is occupied. These sites are supposedly instrumental in phasons that have been studied by various groups, *e.g.* [26,25].

Finally a word about ternaries and quaternaries that we have not discussed because of their added complication. Much research has evolved that describes a wide variation of properties. Among these are morphology and growth kinetics. Some of the most striking results, recorded with synchrotron X-rays, are of faceted growth of quasicrystals from the melt in AlPdMn [xiv] .

8.3 The hierarchic method:

-The premise for our method is that composition is critical to the formation of quasicrystals.

-In any structure, stoichiometry is therefore fundamental.

-The structure must be instrumental in producing icosahedral diffraction patterns.

-The icosahedral unit cell is preferred but not essential.

a. for a unit cell the stoichiometry must be proved.

b. if more than one cell is used, decoration must be supplied and stoichiometry proved.

-Atomic sizes must be consistent with any proposed structure having sufficiently dense packing.

-The method extends short range symmetries to the long range. Type II icosahedra satisfy; poly polyhedra are problematic:

[xiv] http://www.esrf.eu/news/spotlight/spotlight8quasicrystals/ links to a video, recorded using synchrotron X-rays, of faceted quasicrystal growth from a melt of AlPdMn.

Figure 14. Alignment of two icosahedral clusters by sharing of one apical subcluster.

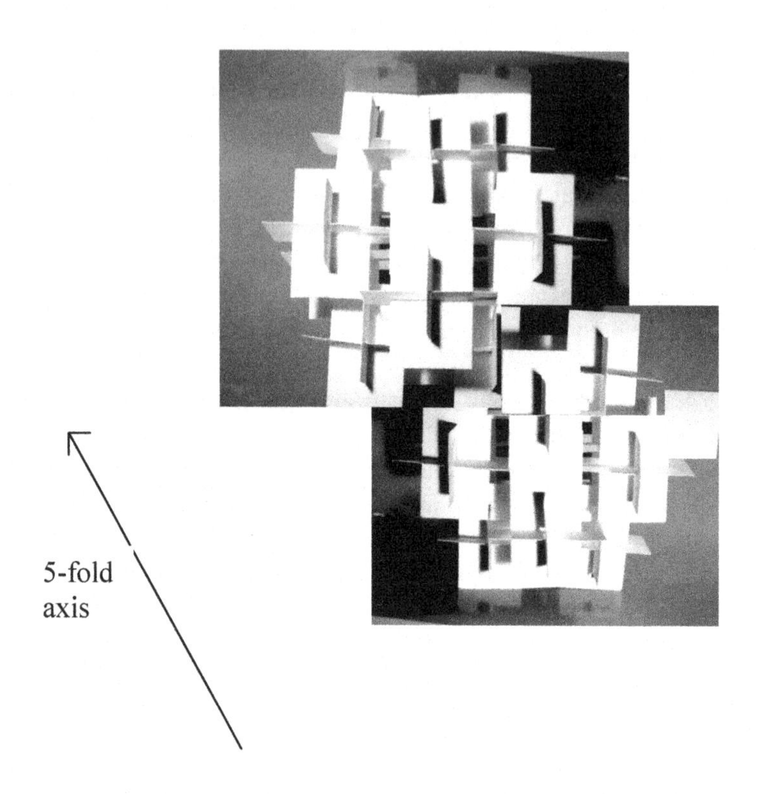

9.Conclusions

The metric for the quasi-Bragg law is discovered, theoretically analyzed for the generic case, and simulated for the hierarchic case. The law has three features: its orders are in geometric series; it has the special metric; and all 'Bragg' orders are forbidden except the second. The analysis shows these features resulting from half integral short range values of the series, combined with integral long range values. The proof is by simulation. The facts that the diffraction pattern is completely indexed and correctly simulated in three dimensions, shows that the simple three dimensional indexation is appropriate. The consistency with experimental data, verifies the method that has enabled the novel measurements reported here. The hierarchic structure is ideal, since micrographs demonstrate frequent defects in the aligned, icosahedral, unit-cell, approximant. Since this is the first measurement of the metric, it follows that it is the first reliable measurement of structure.

Acknowledgement

Nova Science Publishers is acknowledged for their agreement to reprint figure 7 and parts of the text of section 2.

Chapter 3

Evaluation

Contents

3.1 Purpose 65
3.2 Review 3 by Section editor 67
3.3 Review 4 86
3.4 Reviews 1 and 2 91
3.5 Review 5 102
3.6 Review 6 105
3.7 Conclusions 107
3.8 Epitaph 108

3.1 Purpose

A debate is described with referees of two journals, one of which had invited a review. The purpose is to help the reader evaluate the primary document [xv]. This chapter connects six reviews with responses and forewords. They are presented, not in chronological order, but in order of significance. The chronological order is shown in table 3.1. It serves as a list of contents for the chapter. For example, the third review was written by a supposed expert in the field, whom we had complained was ruled out by bias. The reader will judge if he was successful in defending his territory.

It may be surprising that the weak defense they describe is the best they can give. They are all inconsistent mutually and internally. For example, reviewer 4 exalted reviewer 3, but took the opposite view of the Quasi-Bragg law, which however, he misrepresented twice in extreme ways. Do knowledgeable people make weak arguments? There are obvious conclusions. The first is that there is no doubt the metric is now measured for the first time. The Editor in Chief tacitly agreed twice to publication of this debate.

[xv] But see 'steady on' comment on page 8.

Table 3.1 Chronological order of reviews, responses and forewords. Section 3.2. has priority.

History for the review of *The Quasi-Bragg law and metric*	
Anonymous reviewer 1: **3.4.2**	Response : **3.4.3**
Anonymous reviewer 2: **3.4.4**	Response : **3.4.5**
1st appeal to Editor in Chief (EC) : **3.4.6** , objecting to bias in Section Editor (SE)	
rejected	
Appeal to SE **3.2.2**	**Response : 3.2.1 and 2nd appeal to EC**
Following 2nd appeal to EC Anonymous 4th review! **3.3.1** EC closes unresolved case	Brief response: **3.3.2**
First anonymous review from *Materials* **3.5**	Unacknowledged brief rebuttal
Second anonymous review from *Materials* **3.6**	Unacknowledged brief rebuttal
	Summary conclusions : **3.7**

3.2. Review 3

Firstly, response to (3.2.2), appeal to the Section Editor (SE). This has virtually nothing in common with review 1 (3.4.2), review 2 (A4.4) or review 4 (3.3.1)

3.2.1 Response and 2nd appeal to Editor in Chief (EC)

Dear Professor Kostorz,

wo5007 appeal

Following the procedures of IUCr, I appeal against the reviewing of my paper. I ask you to reinstate the advice of reviewer 1:

> "I support its publication in a high profile journal such as Acta Cryst. A."

I attach the review of your Section Editor (SE). For reasons outlined in my two previous emails to you, this is a difficult paper for him to review, I now fear you expected too much.

I have discovered the Quasi-Bragg law. It has a special metric, and I have measured it.

Both I and he claim to be conventional, but we are conventional in different ways and view different timescales. The title of the paper is "The quasi-Bragg law and metric…" and this is where we are furthest apart.

Details are examined in the appendix (following); in this letter I summarize conclusions. I cover his criticism under three headings, and I am able easily to respond to all of them.

1. He claims I do not identify atomic positions (para 2)
2. On novelty,he claims:
    2.1. my quasi-Bragg law is identical to his Bragg law (paras 3-4)
    2.2 his PAS is identical to my metric
3. He claims to observe a different set of conventions.
4. He has impressions

I respond with some factual observations and a discussion of logical requirements in science.

1. I *measure and calculate* the precise atomic position and specie of each atom in both an infinite ideal icosahedral solid, and in a micrograph of a thin specimen observed at optimum conditions[xvi]. Measurements are consistent with known atomic sizes, including the unit cell and with diffraction patterns.

SE writes, "one has to identify atomic positions and refine their coordinates." I have already done this. He thinks inside a box. All that his methods can teach is something about defects and glassiness: not general science, but laboratory preparation. It is important for us not to repeat the error that Pauling made (cited in paper) when he neglected transmission electron microscope data.

In SE's convention the dimensions are chosen:

> "...we just have to choose d_H of the respective scaled PAS[xvii] properly."

This is guesswork. The difference on its own justifies publication of my paper, but there is more to come.

2.1 Bragg's law versus the quasi-Bragg law.

SE wrongly claims that his formula is "nothing else than your quasi-Bragg law." In the appendix, I give the evidence why this cannot be the case. In summary:

- 2.1.1 the Bragg law contains orders that are not observed in quasicrystals;
- 2.1.2 the orders are in geometric series as described in the quasi-Bragg law;
- 2.1.3 simulations show that the diffraction is second order Bragg as in the quasi-Bragg law. (see heuristic derivation in appendix)
- 2.1.4 The geometric series has properties that explain the simulation.
- 2.1.5 The quasi-Bragg law contains a measured metric examined further below.
- 2.1.6 The quasi-Bragg angle is different from the Bragg angle

---

[xvi] The micrograph is given in Bourdillon2009b and can be viewed most easily on http://www.youtube.com/watch?v=0LDS0sQQvpk

[xvii] Periodic Average Structure. He claims this as a chosen equivalent of my 'metric'. This is a guess, described below.

2.1.7 He even states that it makes no sense to apply the Bragg law at all, that is before he claims to apply it.

Item 2.1.2 is especially significant because it enables us to index the diffraction pattern in three dimensions (3D), and to simulate a 3D structure. We are able to do this because the patterns are axial (not lattice[xviii] like as in the Bragg law). Intermediate beams are indexed by vector methods that are standard in electron microscopy.

2.2 'PAS' versus 'metric'

SE claims his 'PAS' to be a conventional form of 'metric', however

2.2.1 His PAS is chosen, my metric is measured.
2.2.2 He does not give the value that is chosen.
2.2.3 It is not mentioned in his Youtube lecture.
2.2.4 It is not given in the proceedings of the 12th International Conference on quasi crystals[xix].
2.2.5 It is not mentioned by reviewer 1.
2.2.6 It "troubled" reviewer 2, who was certainly not familiar with it.
2.2.7 It is not mentioned in structural studies by other crystallographers[xx].
2.2.8 The concept is part of a multiply confused version of Bragg's law that I discuss in the appendix and summarized above.
2.2.9 It is not sensible to equate a generalized *chosen* 'periodic average structure' (in what is elsewhere called an aperiodic solid) with the specifically 'short range' variation that I explain, *measure*, and simulate by (structure factor) means. (As

---

[xviii] as in Hirsch, Howie, Nicholson, Pashley and Whelan, *Electron Microscopy of thin crystals,* Krieger, 1977, appendix 5.
[xix] Phil. Mag. 91 [19-21] (2011)
[xx] e.g. Tsai's group has measured the same value for the 'quasi-lattice parameter' as mine (0.206 nm) and independently use the same index for it, $2/\tau$. It is fortunate that this index is unambiguous and is made without choice (Bourdillon2009b). (see Takakura et al. *Nature Materials* **6** 58-63 (2007).)

explained in the paper, these means are in fact Fourier transforms [equation (1), so much admired by the SE]. The difference is that I succeeded where he failed)
2.1.10 I measure the unit cell dimensions; he can't.

That is why I adapt it. His assumptions appear as a cover for other deficiencies in his own convention as discussed further below.

I was aware of his chosen periodic average structure (PAS) from his review article[xxi]. My measured average occurs in short range (table II in the paper). It is simulated in several different ways (two being illustrated in figure 5 and table I in the paper. Mine is a consequence of the geometric series and of my simulated quasi-Bragg law, which is different from his assumed Bragg law (see appendix and above summary). My metric is so different that it is misleading to call it by the same name.

Meanwhile, no error is found in my working. No fault is found in my theory. On the contrary, he claims his equation

"is nothing else than your 'quasi-Bragg law."

It isn't, as I detail in the appendix. he 'claims' there is nothing "novel" in my paper by misrepresenting its discovery.

3. Convention.

the quasi-Bragg law v Bragg law discussion contains the most important issues, but there are background issues as well. One is my use and precise measurement of the unit cell, i.e. without guesses. By contrast, his proliferation of cells[xxii] is unconventional in the practice of crystallography that has grown over 100 years. I list, in table I below, other divergences between his convention and standard crystallography. With regard to terminology, I make further comment in the appendix.

[xxi] Steurer W. and Deloudi, S., Acta Cryst. A **64** 1-11, (2008)
[xxii] Youtube lecture and review article ref. 5.

Table 3.I. Comparison of three conventions[xxiii]

Physical property	My conventions	classical crystallography	His conventions
Bragg's law	adapted	**yes**	abandoned
No. unit cells	1	**1**	multiple, indefinite
dimensions	3	**3**	6
new crystal definition	no	**no**	yes
driving force/enthalpy described	yes	**yes**	no
atomic size fit	yes	**yes**	not measured
geometric series	explained & simulated	n/a	inaccurate Fibonacci
law simulated	yes	**yes**	no
complete	yes	**yes**	redundant orders
cell contact	double edge sharing	**Face or single edge in silica**	face
phase problem	assumed yes, like silica	**as in silica**	unknown
metric	cs=0.947	1	arbitrary PAS!
metric uniform	yes	**yes**	no
metric explained	yes	**yes**	chosen
metric simulated	yes	**yes**	no
modulus simulated	yes	**yes**	no
long, short range moduli	yes	n/a	no
approximant	hierarchic	n/a	various crystals
structural defects	typically glassy	n/a	unknown (3)
Shechtman's icosahedron	known defective	n/a	ignorant of defects
simulate symmetry error	yes	n/a	no

3. Steurer & Deloudi Acta Cryst A**64** 1-11 (2008)

[xxiii] Table VI in chapter 2 is developed from this early draft

4. Logic.

It is necessary to be sensible. In this instance we have to reflect on the logic of the discussion. There are fallacies that must be avoided. I need to turn your attention to these in order to make sense of conventionality.

The 'informal fallacies'[xxiv] were listed by Aristotle 2400 years ago.

> "the word 'informal' indicates that these fallacies are not simply localized faults or failures in the given propositions (premises and conclusion) of an argument to conform to a standard of semantic correctness (like that of deductive logic), but are misuses of the argument in relation to a context of reasoning or type of dialogue that an arguer is supposed to be engaged in." (*Cambridge Dictionary of Philosophy*, Ed Audi, R.,1999)

The Fallacies have trimmed what is allowed in civilized debate in Western culture generally, and in science in particular since he wrote his *Physics.* The fallacies were endorsed by Locke in the 17th century and have guided modern science.

In this scientific assessment, the only type of argument that is valid is the *argumentum ad judicium* [xxv].

> This "represents a kind of knowledge-based argumentation that is empirical, as opposed to being based on an arguer's personal opinion or viewpoint." Locke compares four types of argument. The first three types come from "my shamefacedness, ignorance or error," whereas the *argumentum ad judicium* "comes from proofs and arguments and light arising from the nature of things themselves"[xxvi].

[xxiv] Cambridge Dictionary of Philosophy, ed R. Audi, CUP 2009. See entries under 'informal fallacies' and under 'Aristotle'.

[xxv] ibid.

[xxvi] ibid.

Since no error has been found in my measurement or theory, my model must be judged genuine. The first reason for publishing is the novelty of the result. However there is a further and related reason beyond its success, namely the intellectual economy of the method. Unnecessary cells and dimensions are slashed by Ockham's razor. Of course some (with long beards!) will call this unconventional, but it is not more unconventional than was Galileo in his time, much less so. In the time span of crystallography, my method is the more conventional. But then, the argument from conventionality is not *ad judicium,* so we should not spend time on it. Discussion must be about the truth, consistency and relevance of the finding. My argument is not for 'quasicrystallographers' (as in review 2) but for blood on bone scientists.

His conclusion that I "should use the scientific language of the field" is vacuous, *ad populum*, and negated by many previous publications from Phil Mag Lett **55** 21-26 (1987) to Solid State Communications **149** 1221-5 (2009), etc. Suppose I write, "I have the impression" he has not read one million papers in crystallography. I have taken "into account" every one of them." My observation is truer than his. But neither is *ad judicium,* and both are fallacious.

Where you find logical fallacy, you should also expect fundamental contradiction. How can something which is in fact new not be unconventional? I summarize my novelty:
-My dimensions are measured; SE's are chosen.
-My law describes geometric series as observed; his describes a false lattice.
-In consequence, I use 3 dimensions; he has to use 6.
-My law is consistently simulated; his is simulated wrong.
-My law is simulated and expounded in second Bragg order; he has assumed first order.
-My atoms are specified, precisely located and measured, both in the model and in the micrographed specimen; Nobody knows "where are (his) atoms," (Per Blek, as quoted in his review).
-I have a unit cell; he has many units (!).

-and he even claims it makes no sense to apply the Bragg law, before he claims to apply it!
-and his 'convention' didn't notice his founding data are not even icosahedral.

He claims wrongly that my paper has no "novel ideas"; and then claims it is unconventional. This is the fallacy called *argumentum ad verecundiam.* His error in understanding leads him to impose a false authority. New wine is not kept in old wineskins. Novel ideas are hidden when dressed in old language. SE describes no term that is unclear. This would be difficult, because the paper is well written.

By contrast, there is nothing in crystallography more unconventional than using 6 dimensions when they are not needed and are indeed cryptifying.

I have shown that my working is novel. My method, as developed from Bragg's law, is perfectly conventional.

Do you not think this debate should be published separately.

Yours sincerely,
Antony Bourdillon MA DPhil(oxon) PhD (cantab)

Appendix for appeal to EC:

Detailed responses to SE's review in order of paragraphs:

Para 1. Cartesian coordinates
What he says is not quite right so I clarify. What I present is a model with a single, tightly bound unit cell. I have used the model to derive and simulate a generic quasi-Bragg law as discussed further below. The dimensions of the model are strictly measured and consistent with known atomic sizes.

Meanwhile, he claims that I 'emphasize' the Cartesian coordinate system. This is a misreading. *All* I say is;

> "Thirdly, when the icosahedron is represented by 'golden triads' the structure can be quantified on Cartesian axes, so that many physical properties are calculated **simply**" (bold type added).

and

> "The equation is **simple** to apply because our axes are Cartesian.

The Cartesian axes are used for convenience and not by emphasis. he is correct, mathematically, that they are relative, but the computational convenience turns out substantive in large calculations on 10^8 individual atoms, each specified, precisely located, and measured for size. What point does he try to make? My methods are simpler and have succeeded where his failed:

Para 2. X-ray crystallography
SE claims:
"It is not sufficient to know that a structure is periodic or quasiperiodic; one has to identify atomic positions and refine their coordinates against a large experimental data set with up to several thousand x-ray reflections (and thousand of unique reflections). This can be done based on tiling models or using the powerful higher-dimensional approach."

He presents a lovely wish for a mathematician or X-ray crystallographer, but:
1. It hasn't been done and he gives no citation.
2. It can't be done for the same reason it couldn't be done in amorphous silica, namely the phase problem.
3. Binary quasicrystals can't be grown sufficiently large and ternaries are more complicated.
4. The quasicrystals are defective, especially the binary ones and certainly those I have studied (as cited). I observe and understand them as glassy and he doesn't know otherwise[xxvii].
5. I have identified every atomic position in an infinitely large ideal quasicrystal, and also in the micrograph that is shown in my previous publications.

[xxvii] As claimed in reference [6] above

6. I have already made the refinement within the limitations appropriate for structures with glassy defects.

What is he asking for? The calculations he wants are regular. His experiment is problematic. Electron microscopy, like Shechtman's, is more readily interpreted. I already have sufficient evidence. The structure of quasicrystals does not have to be a branch of mathematics. Evidently, his non-descript hyperspatial attempts are not as "powerful" as my realistic and measured detail. The paper is about diffraction and he has turned to structure which is not here complete because the quasicrystals are glassy. This is called the fallacy of *ignoratio elenchi*, failing to keep to the point.

Heuristic narrative as introduction to paras 3-5

Naturally, as I was constructing the model, I was thinking about how to simulate the diffraction pattern. Understanding the relation of Bragg's law to structure factors, I was able to arrive at the pattern by what I at first expected to be straightforward means. The version of Bragg's law was the same as SE's except that d_H in first order *i.e.* with $d=a/h$, a being the edge length of the unit cell, and h the index, derived unambiguously by prior means cited in the paper. To my surprise the structure factors were all low and haphazard. I knew what values to expect because it depends on the number of scattering atoms. By scanning the Bragg angle, I found the quasi Bragg angle, θ' and corresponding interplanar spacing d'. I then set about checking that the diffraction pattern was indeed logarithmic (or geometric or [restricted] Fibonacci) and that the second order Bragg (n=2) was zero because higher orders are not (with exceptions) observed in the diffraction patterns. To my consternation the structure factors increased by orders of magnitude in second order and it took me a long time to work out why. I found moreover that d'/d was the same for all peaks. I did simulations on imaginary systems that are different in long and short range (summarized in table IIa in my paper) and found that the reason for the compromise spacing effect (or d'/d or the metric or whatever) is due to the contrast in properties of the geometric series in long and short range. I could predict and measure *d'* directly and relate it to the structure. If this experience is not "novel", somebody is not thinking but being wishful.

The examination of the geometric series gave the explanation. The result is my quasi-Bragg law. You can see that SE is wrong in his assessment. My law is more complex than his and mine is consistently simulated. With this law, coupled to my diffraction theory and 3-D indexation, I measure all dimensions in the unit cell. They are consistent with atomic sizes, and in some micrographs I locate every atom. Given what he wrote in his review (quoted above), most of this is "novel" and of considerable importance. This work began when his law proved false.

<u>Paras 3-4. Bragg's law</u>

He rambles over his Bragg law, pretending similarity. My Quasi-Bragg law:

$$\lambda = d'\tau^{-m}\sin(\theta')$$

is different from his law:

$n\lambda = 2d_H \sin(\theta)$,

because mine is in Bragg second order with measured metric and incorporates the geometric series. The details and explanation are given in my paper. My equation is simulated by individual scattering from 10^8 atoms. My *d'* is predicted and measured, unlike his "chosen" $d_{H,}$ which has lost the significance found in crystals. Moreover, his high Bragg orders are not found in quasicrystals except by exception.

Since the difference has not been understood, I have to illustrate. I index in three dimensions to do justice to the geometric orders and I can do this because my basic system is axial. (Off-axis, see the indexation of axial patterns in my paper - they use the conventions common in electron microscopy.) As an example, in the quasicrystal, the axial indexation on the 5-fold pattern along a 2-fold axis progresses:

$(0,0,0)\ldots(2/\tau^2,0,0),(2/\tau,0,0),(2,0,0),(2\tau,0,0),(2\tau^2,0,0)\ldots$.

This compares with Bragg orders in the fcc Al matrix in 2 phase Al_6Mn oriented on the [0,0,1] axis and progressing towards [1,0,0]: (0,0,0),(2,0,0),(4,0,0),(6,0,0),(8,0,0)....
The latter works for his law. In the former, my three dimensional indexation is descriptive and simplifying for quasicrystals.

He writes: "we obtain $n\lambda = 2/(k\tau a^*)\sin\theta$, which is nothing else than your "quasi-Bragg law." It is not, as you can very well see. The order *n* is not a variable, but has (with a noted exception) the value 2. High Bragg orders are forbidden in the diffraction pattern. My simulations on the hierarchic model showed that his law is wrong. I explain mine much better, as you can read from the heuristic description above, or in my paper. His law is different from mine in essential ways. Translating to his terms in one dimension, the simulated law is:

$$\lambda = \tau^{-m}/(c_s k\tau a^*)\sin(\theta')$$

where c_s is calculated, θ' is the quasi-Bragg angle, and in icosahedral units $a^*=1$, the inverse of the edge length, which is also the diameter of Al in Al_6Mn. For the single case of first (geometric) order he can write $m=0$, and make his $k=1/2$ for the third term of the geometric series illustrated in the previous paragraph; but *n* is discarded, c_s is measured, and the equation is radically different. These considerations are implied in table IIb where the quasi-Bragg law is clearly "novel" and the discrepancies are obvious.

However, besides being wrong by simulation, notice that "nothing else than" is not enough: He has then to go on to show why he needs the complexity of hyperspace, which had not before been used in crystallography. He needs to show that he needs an indefinite number of 'unit' cells. This contradiction is not the way of science. He needs to know his crystal is not defective. He needs to show economy of both understanding and representation in his system, and whether he works on binary, ternary or whatever quasicrystal, he needs to locate individual atomic species. Conventionality is a fallacy, but it is actually on the side of the logarithmically periodic solid

(LPS). My advance has been made because of adherence to the diffraction pattern, to conventional dimensionality, and the other methods of classical crystallography.

Moreover, what can be proved with the wrong law of diffraction? He writes "It does not make sense to apply (the Bragg law) to non-periodic structures and it is not needed for structural analysis." But he then goes on to try to make sense of it by giving a formula which he wrongly claims to be "nothing else than your quasi-Bragg law." On the contrary, major theoretical advances follow the correct adaptation of the Bragg law, including the return to 3 dimensions.

What is he talking about? First it makes no sense, then it makes sense. His review is not thought out. He has not given it enough time. He claims to do crystallography without Bragg's law and has proceeded to multidimensional confusion. He then cannot recognize the adaptation of Bragg's law. (Elsewhere I show how Bloch waves are adapted to icosahedral structures and give reference in the paper introduction).

Para 5. Terminology

By the way, the Pisot-Vijayaraghavan property is a fancy name that I am familiar with. However that is not the point, which is the comparison between long range and short range, including the moduli and metric. The effect has been simulated in the earlier papers cited, as indicated in table IIa. All of the properties of Fibonacci series that we use are well known in mathematics as I have repeatedly explained. What has not been realized is how they come together in the LPS simulations and determine measurement by the quasi-Bragg law. My simulations (figure 5) indeed illustrate that Pisot-V… property but in the sense modified in the paragraph preceding table IIb in the paper.

With regard to terminology, the model has been so productive that it was necessary to coin new terms on the way. The first was the 'golden triad' because, though the structure is well known in mathematics (Huntley R.E., *The divine Proportion* 1970) and though the 'golden rectangle' is commonplace, there was no recognized name for this much used unit structure. The 'logarithmic periodicity' also contains a complex of meanings that I have discussed in previous publications (hierarchical, infinitely extensive, uniquely aligned, uniquely icosahedral, with logarithmic indices due to the geometric series etc.) It is contained in the title of a previously refereed paper. Then there is the 'metric' calculated for the first time, not guessed like his PAS. Because of the model, SE's 'Fibonacci' sequence proved inaccurate, since the geometric series is a special case with ratio between sequential terms equal to τ *exactly*. If he finds this terminology unconventional there are very good reasons for it. Then, by simulations of structure factors, I found other "novel" features (e.g. quads and triplets) that I had to designate...and so on. Moreover, because of the geometric series, I was able to reduce dimensions to the normal 3 for properly descriptive indexation. This was convenient for the calculations. I could labor this point further without overstating it.

If there have been ten thousand publications in quasicrystals, there have been a million in crystallography. The *argumentum ad populum* is a fallacy. Only *argumenta ad judicium* are sufficient. I do not claim incompatibility with hyperspace. Only, it is unnecessary and is struck out by Occham's razor according to normal scientific practice. What is more unconventional in classical crystallography than to invent dimensions that are not needed? However, it is not the way of science to indulge controversy, and that is why I say nothing in my paper about those methods, which are raised only by reviewers. It is difficult to give up cherished ideas. I do not encourage inaccurate measurements. There is however a more tolerant approach: in important branches of physics, rival conventions are enjoyed, as in the Schroedinger and Heisenberg interpretations of quantum mechanics, so two conventions each have, within reason, their place.

I summarize bluntly a few of several advances in terminology that I have adopted:

-His chosen 'PAS' is not equivalent to my measured 'metric'.
-His 'Fibonacci sequence' is an inaccurate version of my 'geometric series'.
-His Bragg law is wrong. It is not equivalent to my quasi-Bragg law.
-My geometric series enables a conventional 3 dimensional indexation and description.
-The diffraction pattern is axial; not lattice-like.
-Tiling is economically described on closed dodecahedral surfaces; not planar surfaces.

Is it not obvious that his criticism of 'hidden terminology' is a cover for disputed theory? If only he could see that his preference for 12-fold coordination (Youtube lecture) leads directly to the unit cell, the LPS, the quasi-Bragg law, and the measured metric.

Para. 6. Ad populum
SE's 'impression' is false. He cannot know how wide my reading in fact is. It is not necessary to cite work proved wrong. The most iconic paper in physics has no references (Einstein, A. 1905 Ann. der Physik[xxviii] **17**, 1905). It is not my purpose to instruct closed minds, and if it is the duty of IUCr to reinforce old myths *ad populum* then its logic is faulty and I hope it is not true.

Your three reviews are neither internally consistent nor consistent with each other.

3.2.2 Review 3 from Section Editor

The following was written before the above response. The demolition is unhappily forced.

[xxviii] Einstein, A., *Ann der Physik* 17 (1905)

Dear Dr. Bourdillon,

I read your manuscript wo5007 with interest and tried to identify novel ideas hidden in your somewhat unusual terminology. If I understand your manuscript correctly, you present a structure model based on superclusters of icosahedra (triads). You also emphasizes the importance to stick to a Cartesian coordinate system. Of course, you know that it does not matter on which basis you describe your structure, this is just a matter of usefulness because you can easily transform different bases into each other.

The structure model you present certainly shows some features known from quasi-periodic order, but real quasicrystals are a little bit more complex. It is not sufficient to know that a structure is periodic or quasiperiodic; one has to identify atomic positions and refine their coordinates against a large experimental data set with up to several hundred thousand X-ray reflections (and thousands of unique reflections). This can be done based on tiling models or using the powerful higher-dimensional approach.

In your manuscript, you address the question of the validity of Bragg's equation $2d_{\mathrm{H}} \sin\theta = n\lambda$ in case of quasicrystals. As you correctly write, the sequence of distances between atomic layers along a given direction is in particular cases related to the Fibonacci sequence. In case of an X-ray beam reflected on such a stack of atomic layers, Bragg's equation cannot be applied if we do not know how to handle the term "$d_{\mathbf{H}}$". However, we know that for every reflection with index **H** and its higher harmonics, with indices $n\mathbf{H}$, there exists a periodic average structure (PAS) with period d_{H}, which has to be used in Bragg's equation. Then it can be applied as usual. The scaling symmetry by powers of τ of the diffraction pattern, which is a Fourier module of rank n ($n > 3$), does not need to enter Bragg's equation; we just have to choose d_{H} of the respective scaled PAS properly. Anyway, Bragg's law is just a visualization of constructive interference from a stack of equally spaced lattice planes. It does not make sense to apply it to non-periodic sequences and it is not needed for straucture analysis.

Since you claim to explain diffraction on quasiperiodic structures by your modified Bragg equation, I want to discuss this a little bit more in detail. Without loss of generality, I will chose a 1D example, the well-known Fibonacci sequence. Its kinematical diffraction pattern M^* can be described by a Fourier module of rank 2:

$M^* = \{H = ha^* + k\tau a^* \mid h,k \in Z\}$, with diffraction vectors H, reciprocal space basis vector a^* and Z the set of integers including zero. For periodic structures we have $H=1/d_H$ and can write Bragg's equation in the form $2d_H \sin\theta = n\lambda = {}^2\!/_H \sin\theta$. If we chose the special case h=0 and consider the scaling symmetry of the Fibonacci sequence (what you call "logarithmic periodicity") expressed by $H' = \tau^n H$ with n, any positive or negative integer. then we obtain $n\lambda = {}^2\!/_{(k\tau a^*)} \sin\theta$, which is nothing else than your "quasi-Bragg law".

By the way, your observation that τ^n aproaches, with increasing n, integer numbers better and better is long known and a necessary (but not sufficient) condition for a pure point Fourier spectrum (Pisot-Vijayaraghavan property).

To summarize, I have the impression that you are not taking into account the more than ten thousand publications on quasicrystals that appeared so far. The crystallography of quasicrystals is already well established and if you want to contribute and to be understood by the community you should use the scientific language of the field.

I am sorry, your manuscript does not meet the necessary requirements for publication in one of the IUCr journals.

Best wishes,

Walter Steurer

Walter Steurer
Section Editor Acta Crystallographica A

3.2.4 Prior appeal to Section Editor

Dear Dr. Steurer,

wo5007

Gernot Kostorz has insisted that I write to you even though I have told him I doubt you are qualified to judge my paper because of your prior intellectual commitment. He has promised to exercise policy if I cannot be satisfied by the result. I take it as advantage that he knows you well enough to trust your judgment. Throughout this letter I shall indent supporting information.

> Since my first paper on quasicystals (published in 1987 and cited in my current paper) I have taken the alternative view outlined by Bursill and Peng (also cited). "*What is most intriguing, of course, is whether we are concerned with a material having singular structural properties because of the chemistry of Al and Mn, or whether the principles suggested by the quasicrystal concept will find more widespread application*". With the aid of a hierarchic model, I have found the solution for the underlying law of diffraction for quasicrystals. I know, from my own experiments, that the hierarchic model is defective. That is why my paper is principally about diffraction; not structure.

Nevertheless, the diffraction principles apply generically because of their derivation.

Referees do not like to hear that the structure is like silica glass; though none so far have said why not. The chief difference is that here the icosahedral units share two edges instead of one on silica tetrahedra. The dual edges cause the icosahedra to align. The geometric series in the diffraction is a natural consequence, as I prove for the hierarchic case.

I will

1. outline my discovery
2. explain what is wrong with the previous Acta Cryst. reviews
3. ask you to re-evaluate my paper while I offer to provide what help I can.

1. I have discovered the quasi-Bragg law. It has a special metric. The discovery is due to the application of the hierarchic model that I describe.

For an introduction, you may view my video on Youtube: http://www.youtube.com/watch?v=0LDS0sQQvpk. It was not available until recently. It is constructed for the general view and is not tightly written as is my paper. The video opposes the view taken in your lecture on Youtube.

The discovery is backed by simulations that are both *internally* consistent, and consistent with *experimental data*, including the micrographs observed at optimum defocus. Indeed the discovery arose from those simulations which at first seemed inconsistent, but are now proved fundamental.

The discovery was made possible by my cluster model with golden triads. Coordinates of all atomic positions are easily identified on Cartesian axes. A range of calculations could then be performed.

> The data are backed by a significant body of development that has been published spasmodically, for the reason already given. One referee was extremely positive. I think you need to be aware of the background, in order to avoid short cuts.

2. Your own two referees for Acta Cryst. expressed opposite opinions: reviewer 1 thought the paper should be published in Acta Cryst.; the other thought not. The subeditor rudely wrote that the paper is "flawed" without giving his reason. I give my rebuttals in an attachment. The attached paper contains minor emendation, though I have also changed the title because conciliation has not worked.

Notice that neither referee has summarized the main findings and significance of the paper by reference to its title. I believe this is because neither has understood it, for various reasons. I think the new video will help you do so.

In particular, reviewer 2 objected to looking up the references and apparently failed to do so, presumably because he did not allow time for it. He was therefore at a singular disadvantage. In my rebuttals, I have tried to show that all necessary information is contained in the paper. It is impossible to put the whole supporting evidence in a single paper within journal space limitations, even if a journal would allow it. However, to ease referencing, I now add to my list of references, several supporting pages easily available on the worldwide web.

Every factual observation made by both referees is wrong as I demonstrate in my rebuttals.

> Moreover reviewer 2 is inept, as I show at several points. He does you no justice by calling your expertise 'quasicrystallography'.

3. The fact that the references are bad does not turn them into a recommendation, so I have to ask you to re-evaluate my paper. The discovery is

fundamental. The special diffraction mechanism is described in detail and proved by simulation. I would be happy to clarify any doubts that you may have, and if there is any way I can improve the paper I would welcome your help. Your time is appreciated and I trust you value your role accordingly.

Finally, I think you should follow the advice of reviewer 1: "I support its publication in a high profile journal such as Acta Cryst. A." I agree even though he clearly hasn't understood that the paper is about diffraction which has wider application than the discovery model, itself. Should you need time to be convinced by my arguments, are you sure it would not be short sighted to discourage debate? I look forward to seeing others corroborate my findings.

Yours sincerely,
Antony Bourdillon, MA DPhil (oxon) PhD (cantab)

3.3 Review 4

3.3.1 Preliminary guide

Whether its author knows it or not, the following report is a multiple lie.

3.3.1.1. He asserts our indexation is incomplete,

> 3.3.1.1.1 Which beam in the Shechtman et al data is not both indexed and simulated? He is a dreamer.
>
> 3.3.1.1.2 He gives no example. Reader should assume he can't since his omission is a breach of the most basic rule in science, that every statement must be justified.
>
> 3.3.1.1.3 No reputable journal allows such unjustified opinions to be expressed in any of its papers.

3.3.1.2 Yet every diffracted beam about the three major axes is both indexed and simulated,

3.3.1.3 The 'sound demonstration' is evidential and theoretic.

3.3.1.4. Which is complete? *His* theory without a metric; or *our* complete simulation and measurement.

3.3.1.5 The report is not consistent with the tradition of IUCr. This referee "cannot find any new result" because his theory and metric are imaginary. The reader who rejects the report *in toto* loses nothing.

3.3.1.6 To summarize: *in science measurement rules; indexation is a convention; and the metric is measured.*

I measure the metric for the first time where he "could not find any new result.." Choice of indexation is independent of the measurement of the invariant metric. Why has he not measured the metric? Why can't he do so with his own imaginary indexation? He must read the title because missing the point is a logical fallacy.

The quasi-Bragg law is nakedly simple. His formulae have confused even himself and his transcription is inaccurate. Review 3 claims the two laws are the same; report 4 accepts differences in the laws. His claim to support SE is therefore obfuscating. Neither has measured the metric.
Show us the goods.

Our paper describes and demonstrates how the law is applied – in the same way as Bragg's law. Since he fails to "find a clear and sound demonstration of the formula", here it is:

a. It is the same as for Bragg's law , except that:
b. The orders are obviously and evidently logarithmic, and simulated;
c. The diffraction is in second order Bragg because of the half integral values in the geometric series, as simulated;
d. The quasi Bragg angles and quasi interplanar spacings are subject to a special metric (acknowledged by Section Editor), here measured for the first time.

He claims that, "The author's formula takes account for [sic] only a fraction of the possible diffraction peaks". *Not for the first time, the referees of Acta Crystallographica have proved incapable of following the most basic rule of practical science: every statement must be supported by evidence.* The Editor knew of the failure because of the previous supplementary information submitted by his Section Editor after criticism. *What to expect from referees in denial? These are supporters of inverted 'scientists' that 'indexed' and 'simulated' a false pattern, 'without decoration', and who don't know what they were doing.*

Their methodology has led to false logic: we index completely in 3 dimensions (based on a tetragonal subgroup of the icosahedral point group symmetry[xxix] in a hierarchic model); calculate the pattern and measure the metric. The referee thinks that the indexation is not complete because it was previously done in 6 dimensions. The fallacy is in formal logic rather than 'informal' [xxx]. Moreover, by reviewer's informal logic, Newton has yet to prove his law of gravity from the epicycles of Ptolemy. The Editor declared (email at end of section 3.4.6) his skepticism with the reviewing process. Pity cardinal Bellarmine. *Vive* Galileo.

As to the references in the Report: notice that we referenced three of them 25 years ago; it is counterproductive to repeat them now. The remaining, subsequent review is a sign that all is not well. Their structures are undecorated; without metric; have a lot of dimensions; and other objections were already recited in section 3.2.1. Is it a mind that calls 6-dimensional hyperspace a "clean and simple demonstration" where they are proved redundant in multiple ways? Referee's citations are not ignored, they are properly not referenced. Unfortunately, Referee's crown jewel is mostly glass, where the double meaning is intended.

By contrast, electron microscopists prefer three dimensions, as in their indexation of the hexagonal close packed structure [xxxi], where X-ray uses four dimensional hyperspace [xxxii]. Our indexation is (1) instrumental in, (2) sufficient and complete for, and (3) descriptive of our purpose, which is the measurement of the metric for the first time. Without an argument, how can he "convince" anyone that knows science is about evidence and deduction, or anyone that values formal and informal logic above conviction and self-justification?

[xxix] Bunker, Philip, Jensen, Per (2008), *Molecular Symmetry and Spectroscopy,* 2nd ed. , Ottawa: NRC Research Press, ISBN 0-660-19628-X
[xxx] The Cambridge Dictionary of Philosophy, ed. R.Audi, 1999
[xxxi] Hirsch, P.; Howie, A.; Nicholson, R.B.; Pashley, D.W.; Whelan, M.J., *Electron Microscopy of thin films*, 2nd ed. 1977, Krieger, NY, appendix 5.
[xxxii] Cullity, B.D., *Elements of X-ray Diffraction,* 2nd ed., Addison-Wesley. US, 1978, ch. 4.

3.3.2 The report

Report on paper wo5007

The paper wO5007 announces the finding of a "new Bragg law" concerning the specific case of icosahedral quasicrystals:

$$\lambda\tau^{m} = d\sin\theta$$

The referee could not find a clear and sound demonstration of this formula and could not find any new result with respect to the usual indexing schemes of Shechtman et al.[1] and Elser[2]. It has been demonstrated there that the reflections can be indexed with 6 indices and that their lengths in $\mathbf{E}_{\|}$ and $\mathbf{E}_{\perp}$ are given by:

$$Q_{\|} = \sqrt{N + M\tau}/K \quad Q_{\perp} = \sqrt{N(N\tau - M)}/K;$$

where N and M are integers and $K = A\sqrt{2(2+\tau)}$; where A is the 6D lattice parameter. This leads thus to the Bragg law:

$$\lambda\sqrt{N + M\tau} = 2K\sin\theta$$

Because $Q_{\perp}$ is the argument of the perpendicular form factor that has its maximum value at $|Q_{\perp}| = 0$, the intense peaks will be among those corresponding the integers N and M being equal the successive Fibonacci numbers $N = f_k, M = f_{k+1}$ in which case $\sqrt{N + M\tau} = \tau^{(k+1)/2}$. This includes, for odd k orders, *the author's formula that takes account for only a fraction of the possible diffraction peaks.* In particular, the diffraction pattern being a $\mathbb{Z}$-module of rank 6, all 3D projections of 6D rational planes, are dense sets of reflexions, not only the 2-fold orientation. The 6D indexation is fully coherent in any of the diffraction planes including the 2-fold (The 2-fold initially

[1] J. W. Cahn, D. Shechtman, D. Gratias, J. Mater. Res. 1, 13—26, 1986
[2] V. Elser, Phys. Rev B, 32 -8, 4892–4898 (1985)

Based on the two referees' reports, it was considered that the manuscript was too flawed.
I have attached copies of the referee report(s).
Best wishes
Professor P. R. Willmott

Dear Professor Willmott,

I have posted the paper on open source without change because, from internal evidence, your reports are questionable.

If you read the abstract you will see that they have nothing to say on the topic. The discovery of, and explanation for, the new metric is significant for general physics beyond crystallography.

The status is:
1. I posit a hypothetical model in real space.
2. I show, by simulation with explanation, that if the model diffracts it does so by a new quasi-Bragg law with a new metric.
3. I show that the simulated diffraction pattern is the same as Shechtman's data after correction.
4. I show that the model is not the final structural solution, but understand the quasi-Bragg law and metric are generic.
5. You take the view my model is not real because of hyperspace!

... flawed unfortunately because its metric is wrong.

Again from internal evidence, reviewer 2 proves incapable of reading anything in context. He is so inept that he even calls himself quasi.

Yours sincerely,
Professor Antony Bourdillon

3.4.2 First referee's report

review from referee 1

Journal: Acta Crystallographica Section A: Foundations of Crystallography
Paper: wo5007
Authors: Antony Bourdillon*
Title: Diffraction of a sinusoidal wave by scatterers in geometric series

The first impression on reading into this paper is that it is "unconventional", primarily because the introductory paragraph is polemical. It goes on to describe a model for an arrangement of icosahedrally-arranged atoms which give a heirarchical structure. This is described verbally, and the words are supported by photographs of what appear to be cardboard models. Personally, I have a real problem in converting the information as presented into something I can handle - I would like to try to follow the rest of the paper, perhaps by making my own FT of the real-space structure - so having a more mathematical description of the atom positions would help. The model is effectively deterministic (eg "the centres of the resulting triplet is (sic) necessarily planar"; "the series repeats indefinitely") so it would help the reader to give a rule for generating atom positions. Much of the following discussion related to "indexing", and so to identifying "interplanar spacings". At best in a non-perioic structure these must correspond to strong features in a 2D projection of the atom-atom correlation function, and indeed, I believe that taking a look at the correlation functions for the type of structure envisaged might be helpful to the reader coming at this proposed interpretation for the first time. Some more care in explanation might also help a reader understand such comments as "In figure 4 the cross is shown to repeat on the third bright ring and again on the diagonal X" - how this relates to figure 4 is not immediately clear, and perhaps it should be?
These various obstacles, coupled with my own shortcomings, make it hard for me to assess the technical reliability of the work.
The author has clearly thought about this material for a long time. I support its publication in a high-profile journal such as Acta Cryst.A. It is for the community to decide if the ideas are useful. I do think the reception of the paper will be improved if the polemical parts are removed, and more attention is given to a mathematical explanation of the suggested structure. Correlation functions for the structure may aid the discussion of "planes" involved in diffraction.
Finally, I believe that readers would find it useful to have a wider range of references available when reading the paper.

3.4.3 Response to reviewer 1 (r1).

I consider his general comments first. They are described in his second paragraph.
(r1) claims that the introduction is polemical. Unfortunately he gives no instance or example. The text should have been taken as written, without hidden meaning. I am glad the referee supported the paper.

Details:
1. Atom sites.
(r1) demands locations for atomic sites. They are given in the paragraph following equation 1. They are simple.

2. Determinism.
Three points always locate a plane. Three points are necessarily planar. The centres of three triplets are likewise necessarily planar. The necessity is from mathematical tautology.
3. Correlation functions.
He should certainly calculate correlation functions though I have already done this.
4. I could certainly fill out an explanation for the cross in (figure 6). However the quickest information comes from staring at the diagram. It was a revelation when the structure was first noticed and even more so when plotted exactly on computer graphics. Indexation, after that, was elementary.

3.4.4 Second referee's report and rebuttal

review from referee 2

Journal: Acta Crystallographica Section A: Foundations of Crystallography
Paper: wo5007
Authors: Antony Bourdillon*
Title: Diffraction of a sinusoidal wave by scatterers in geometric series

In view of the recent Nobel prize awarded to Dan Shechtman, who is much cited in this paper, it is important to consider the material very thoroughly. My major problem with the work is that it is not placed in context with the accepted and well-established field of hyperspace crystallography. In essence it avoids all conventional means of scientific discussion which makes it very difficult to understand what points are new and what the author feels is wrong about previous work. For example the paper begins with "Quasicrystals were discovered in 1982 (Shechtman et al.,1984). Since then, the structure has
been unknown and disputed (Proc. 11th Int. Conf. Quasi Crystals, 2010)." Well this is clearly not the case as there are hundreds of papers reporting structural studies. Some things are known, others unknown. The author must digest what is unknown and add to that body of work, not present a theory in isolation.

There are other troubling statements. Page 3 "For interest, both conditions are consistent with data (Shechtman et al. 1984)" and then a citation to a book chapter is given (Bourdillon 2009b, appendix C), and similarly "A cluster is illustrated in figure 2. We have called it logarithmically periodic'(Bourdillon
2009a)" and again it is not accessible to readers of the paper without finding the books. In general scientific papers should be self-contained, not requiring books to be found.

P5: "It is obvious that the arithmetic orders in
the Bragg law $n\lambda=2d\sin(\theta)$ are, in logarithmic periodicity, converted to indices, $\tau^{-m}\lambda=2d\sin(\theta)$.
However, with such a radical change, it is no longer clear that d is a recognizable interplanar
spacing, and it needs to be debated whether the modified law applies at all when unit cells are
not arithmetically periodic as in crystals having a Bravais lattice. " What does the author mean by a recognizable interplanar spacing, or a modified law? It is not necessary to see any planes to have coherent diffraction. This statement is troublesome, and it is not clear what point the author is making.

Same page: "By contrast in quasicrystals, the brightness of diffraction patterns typically
increase axially to about the third bright ring' (e.g. Bourdillon 1987), and then fall away without
excitation of higher order zones." Is the author suggesting that the Ewald sphere does not curve when studying quasicrystals? There have to be higher order zones, even if the spot intensities are not as simple as in regular crystals. This is not a very scientific statement.

Caption of Fig 3c: "The misaligned planes do not diffract." What is meant by this statement? Atoms scatter whether aligned or not, and constructive interference will occur from any periodic or quasiperiodic structure in particular directions.

p10: "We call the offset the compromise spacing effect (CSE). This effect is
illustrated in Bourdillon 2011. The offset has implications for both the quasi Bragg angle and for
the measured interplanar spacing: d'=dBragg/cs ." This is not intelligible without explanation or access to the book chapter.

p15. "The first is: with which of
the two inconsistent patterns published by Shechtman et al., (1984) should simulations be
compared? The second is how to index them?". More discussion is needed on why the Shechtman patterns are inconsistent.

In summary this lacks proper scientific context in light of the established methods of quasicrystallography which makes it difficult to follow and difficult to understand what point the author is trying to make. The conclusion "The regular LPS is most valuable
for understanding mechanisms in quasi Bragg diffraction" is not substantiated by the body of the paper. I have to conclude that this work will not add to the understanding of quasicrystals unless put in context of prior work and accepted concepts in the field of hyperspace crystallography.

3.4.5 Response to second reviewer (r2):

The paragraphs refer to his review:
Para 1. Context.
(r2) writes that the work is "not placed in context." He mentions hyperspace.
As I wrote in my introduction, "In circumstances where many models have been debated, we attempt, as an exercise in *conventional terms,* a theoretical description of the peculiar diffraction." (bold added) . My paper is not hyper at all; it is about diffraction in real space. My indexation is not hyper and it produces remarkable results. *Since my conventional explanation is valid, hyperspace, which is not conventional in crystallography, is excessively complicated.*

(r2) claims that the structure is known "as there are hundreds of studies reporting structural studies". That his logic is faulty is obvious since Galileo. He even admits that "other (things) are not known" !! (a minimalist understatement especially with regard to 'Bragg' diffraction, as I show). I bring to your attention the common complaint by crystallographers repeated in my covering letter to Professor Kostorz, "Where are the atoms? (http://www.mendeley.com/research/structures-quasicrystals-atoms/)" and to a statement of some time ago but relatively recent since the discovery of quasicrystals, "What is a quasicrystal? The long answer is no one is sure." (Senechal, M., *What is a quasicrystal? Notices to the American Mathematical Society*, %3 886-887 (2006)) These remarks are typical and contrary to the belief of (r2). It *is* the case that the structure is unknown and no amount of believing will change it. *However the paper is not about structure as I make clear in the introduction; but about diffraction, the forerunner to structural studies.*

Para 2. References
r2 complains about two statements that trouble him:

the first one shouldn't because the preceding paragraph is perfectly clear:

> "Two immediate conditions for this structure are:
> " -firstly, if the central atom is called a 'solute' (as in the melt before crystallization), then the ratio of the diameters of the solute atoms to the solvent atoms (here having unit length like the icosahedral edge) is $\sqrt{\tau^2+1}-1$;
> "-secondly the stoichiometry of solute to solvent is 1:6, after accounting for shared boundary atoms."

His second complaint is about a reference to a readily accessible journal article, Bourdillon2009a. He has not checked his facts.

He is troubled by what is really nothing. Thirdly, (r2) complains, "In general, scientific papers should be self-contained, not requiring books to be found." I agree absolutely with his sentiment but the present case is particular. His objection to Bourdillon 2009b is the reason why I offered you free copies of the books. They are catalogued in Cambridge University Library. You would find, in the books many critiques of the refereeing process. There are also laws of copyright. Moreover there is a page limit on journal articles that makes references to books sometimes unavoidable, including the present case. He implies that books are never referenced! However, contrary to his mistaken opinion, the paper *is* self-contained within the objectives stated, and the references are all to secondary material or to easily accessible electron microscope images. The only justified exception is the reference to the copyright material of Bursill and Peng.

Para 3. Meanings of interplanar spacing

(r2) cites

> "It is obvious that the arithmetic orders in the Bragg law $n\lambda = 2d\sin(\theta)$ are, in logarithmic periodicity, converted to indices, $\tau^m\lambda = 2d\sin(\theta)$. However, with such a radical change, it is no longer clear that *d* is a recognizable interplanar spacing, and a debate is needed about how a modified law applies when unit cells are not arithmetically periodic as in crystals having a Bravais lattice."

while I add the context:

> "In the following text we will discover another change when we put n=2. Simulations of atomic scattering from extensive solids will answer the uncertainty. Meanwhile intervening

> points in the reciprocal lattice plane of the diffraction pattern can be filled by vector arithmetic on the indexed logarithmic radials, as is usual in electron microscopy."

He asks "what does the author mean...by recognizable interplanar spacing or a modified law?" and then displays his confusion:
The obvious answer is: the interplanar spacing is either *d* in Bragg's law which has to be modified in the quasi-Bragg law or the related planes in the crystal structure which are to be measured by it. The law has to be modified a second time to include the metric. These modifications are the main point of the paper and the meanings are so clear they should not require repetition because the context copied above points the way.

To know precisely what is meant, he must study the deviation from Bragg diffraction found in figure 5 and elsewhere. In section 4 I go on to explain what exactly is meant by interplanar spacing $d'=d/c_s$.. The chief novelty of the paper is the explanation for c_s based on approximations in the geometric series. Moreover the exponential order is key to the indexation which proved vital to the demonstration. This is the product of the unique model and it is all clearly explained for those with enough time to read it..

Para 4. Ewald sphere
(r2) asks whether I am "suggesting that the Ewald sphere does not curve when studying quasicrystals?' The opposite is true as you will see from the context:

> "In electron diffraction generally, the deviation parameter on the Ewald sphere construction causes near axial reflections to be bright, the brightness falling with increasing deviation parameter until the occurrence of the first order Laue zone. By contrast in quasicrystals, the brightness of diffraction patterns typically increase axially to about the third 'bright ring' (e.g. Bourdillon 1987), and then fall away without excitation of higher order zones."

I write the opposite of what he suggests. All of the diffracted beams that we are simulating fall within the first Laue zone as understood. If (r2) will look at the cited data he will see there are no high order Laue zones in the normal sense. He will also see evidence for the third bright ring, and other publications show the same, providing specimens are thin. Does he not know that amorphous materials also lack high order zone patterns in TEM?

I have considered for some time how the Ewald sphere might intersect the quasi-reciprocal lattice and I make some remarks in the paper. However the quasi-reciprocal lattice is still uncertain so it is not sensible to stretch the concept. It is unlikely to be cubic. It remains interesting that the strongest structure factor is for the third bright ring. As an electron microscopist I also wrote:"

> "Many circumstances affect intensities measured in electron or X-ray diffraction. In electron diffraction, specimen orientation, and consequent extinction distance are critical. Multiple scattering, temperature, specimen purity, beam uniformity, lens aberrations and other factors also influence measurements. If a specimen is sufficiently thin, the diffraction may be studied in the kinematic rather than dynamic approximation, and this is supposed in the following calculations."

These factors are in addition to the Ewald sphere already mentioned. Our purpose is to compare calculated structure factors from an infinite, aligned, icosahedral model, with measured intensities when the specimen is thin. If there are high order zone patterns, they are not evident and that is the reason for the remark. No amount of *a priori* debate changes the fact.

Para 5. Forbidden lines

(r2) asks what is meant by, "The misaligned planes do not diffract"? He claims that "atoms scatter whether they are aligned or not, and constructive interference will occur from any periodic or quasiperiodic structure in particular directions." Has he heard of forbidden reflections? In the fcc structure (100) planes do not diffract. What is apparent is that, for example, the $(1/\tau, 1,0)$ reflection is allowed and so is the $(2/\tau,2,0)$; but the $(1,\tau,0)$ is forbidden. The fact is supported by simulation. I have worried a good deal over it. The data shows that diffraction occurs in linear series in the direction shown (normal to the diagonal planes) and not in geometric series.

The meaning is that if you plot lines normal to the planes identified and passing through the atoms, then the l-s planes (figure 3c) wiggle from side to side across the lines; while the intracellular planes are aligned without wiggle. This alignment is found on all the other main axes. The wiggle is supposedly the origin of the forbidden lines. It is not necessary then for coherence to be constructive as he insists. As the explanation is a plausible supposition, it is not right to labour the point. (r2) should study the figure.

Para 6. Explanation
(r2) claims the explanation for the CSE "is not intelligible without explanation or access to the book chapter." I give the context for his quote:

> "*The reason for this will become apparent in the next section* as will the reason for the second consequence: the peak is offset from the Bragg condition at step 1. The offset, c_s=0.947, is found by scanning the value of the index used $(\tau/2,0,0) \leq h_i \leq 0.8(\tau/2,0,0)$. We call the offset the compromise spacing effect (CSE). This effect is illustrated in Bourdillon 2011. The offset has implications for both the quasi Bragg angle and for the measured interplanar spacing: $d' = d_{Bragg}/c_s$." (italics added)

See the answer to para. 3

Para 7. Shechtman's inconsistency
(r2) says that "More discussion is needed on why the Shechtman patterns are inconsistent." The discussion is given in section 6, last paragraph. Furthermore, based on our simulation, we predict how it should be corrected. Further supplementary illustration is given in Bourdillon2009b, appendix D, though this is illustrative and secondary but doesn't help someone who won't read it.

Why do all of his objections result from failing to read the script in context? Does his schedule allow time for reviewing? **How quasi is it not to notice, after 27 years, that the data are not as claimed ...and double quasi not to recognize it when told...and triple quasi to dispute the fact without reading the references?**

Para 8. Summary
...and ridiculously quasi when he calls his subject "quasicrystallography", without irony but with perfect truth.

3.4.6 First appeal

```
Dear Professor Kostorz
...... .
As our paths have not before crossed I did
introduce myself in my covering letter to you with
```

my original submission. I have discovered the metric for the quasi-Bragg law.

Below you will find an email that I have recently received from Peter Strickland. It misunderstands my appeal, which is not against a decision but against a procedure. **No judge in Europe is allowed to preside over a case in which he has an interest.** Strickland writes that he has spoken to you and this implies that you have not read my appeal. I therefore attach it, below the copy of his email. If you have already read my appeal, I have nothing more to say except that writing to you seems to be in accordance with the published policy of Acta Cryst. I do not consider that Strickland has followed it.

I had already answered, in a previous email to Nicola Ashcroft, the question he asked, so I have not responded to him. As this is a public matter, I expect you want to know what is going on in the office.

Yours sincerely,
Antony Bourdillon

Dear Colleague

Thank you for your mail and your congratulations.

As Peter Strickland wrote, we have to follow our procedures. The term 'procedural conflict of interest' is not clear to me. A scientific conflict of interest may be suspected for any case where an editor is knowledgeable in the field of a paper he/she is handling. However, scientific editing is based on the combined knowledge of (co-) editors and reviewers, and a scientific journal is well advised to have (co-)editors who know their field. The implication that a (co-)editor could be handicapped by knowledge of a field and thus unable to give an author fair treatment sheds considerable doubt on the validity of the currently widely practiced procedures of scientific publishing. Certainly, anyone may have his/her general doubts on these or other procedures, but if one submits a paper, one also accepts the rules of the journal, as applied to all other authors as well. There is no possibility to appeal against procedures written down in the Notes for Authors.

We have our confirmed and very transparent procedures. At this stage, the Section Editor of Acta A is in charge to handle your complaint. If, after his decision, you think you received improper treatment, I am prepared to undertake a final evaluation. This would require a complete documentation and a signed letter pointing out the matters of discontent.

I am very confident that all efforts are being made to give proper attention to your concerns.

Best wishes

Gernot Kostorz
Professor emeritus of Physics
Editor-in-Chief IUCr Journals

3.5 Review 5

Responding to the following two reviews is easy. Where we disagree is the overarching importance of measuring the metric. Anyone that can't see it is called Newtonian or pre-Newtonian. "That is physics. End of story"! - As below.

It is established by the debate in this chapter that ours is the first measurement. There is no debate over the value measured.

As before, no argument is now made that is not false in fact or fallacious in logic [27]. This paper therefore provides the first structural measurement based on diffraction, and it is consistent both with electron microscope calibrations for image magnifications and with known atomic sizes. Neither of the two following referees has a measure for "Where are the atoms?" and neither has understood "Why are they there?" For authors, one good review [7] outweighs an infinite number that are illogical. If journals add any value, it is in the opportunity for debate. The myth is that 'peer review' in a competitive industry can serve science. That is why private publishing is a necessary part. Fortunately in modern electronic times, the method is both easy and searchable. There are more than enough reviews here and in reference [27] to guide any reasonable reader. What is most new is given in the title, the abstract and in table VI.

Notice that of the two following reviewers, one is the "that is physics. End of story" type, followed by paragraphs *non sequntur* (details below); the other is the false hypothesis type, from which nothing follows except that he is a Fibonacci sequence denier, which is extreme. Misunderstanding a small part of an argument is not the same as demolishing it, as they imagine they do.

3.5.1 The report

> "This paper claims to have discovered the "metric" for the quasi-Bragg law for the first time. It claims good agreement between simulations and experiment both in reciprocal and real space. However, it is very hard to discern what exactly is new here. I have watched the Youtube videos from the author, and looked at the books cited, however, there is no enlightenment for me. The author builds models based on icosahedral clusters to compare to experimental data. But it is well known that quasicrystals are based on icosahedral atomic clusters, so the match is certain to occur. What have we learnt from the models? Perhaps the author feels that his quasi Bragg law is new. However, the equations are just Fourier transforms, as they need to be, so again I do not see anything new. Diffraction involves Fourier transforms, that is the physics, end of story.
>
> "The whole work makes sparse reference to the scientific literature. The claimed agreements between simulations and models are due to the fact the author is using icosahedral models. Take the electron microscope image. The only match is the five fold shapes, which are not atoms at all. Again, we knew the symmetry already so what do we learn here? Does the author think that matching the symmetry is enough to locate the atoms? There is image simulation software available on the web, however, the data the author presents are of insufficient resolution to locate the atoms, so it can only be interpreted in terms of symmetry. Again, we learn nothing about the actual atom positions in the real specimen, just that the model does have the right symmetry.

> "I am at a loss to now what to suggest. Scientific work needs to be reported in context of prior work, clearly building on it, and be described such that anyone can repeat the experiment. This does not meet any scientific criteria."

3.5.2 Rebuttal

Paragraph 1: End of story without a metric! Followed by two paragraphs *non-sequntur!* We alone have a single unit cell, measured consistently in many ways. We alone have a uniquely icosahedral structure. Multiple other novelties are summarized in table VI. These facts demonstrate his confusion to be extreme. Moreover the usual soup of poly polyhedra (tetragonal, rhombohedral, triacontahedral etc.) in 6-dimensional space, with flights of the imagination that explain neither why icosahedra form in preference to crystalline or glassy metals, nor why they diffract in icosahedral patterns. Notice that when referees determine to have their own way and can't find anything wrong, they deny novelty. His contention that "so the match is certain to occur" assumes the indexation is correct, as indeed it is, but in contradiction to other reviewers. He does not need to "discern", only to overcome his blindness.

Paragraph 2: It is not true that "the only match is the five fold shapes". Not only the shapes, but also the dimensions consistently match at all levels, including relationships of images to diffraction patterns and atomic sizes. This is all entirely new since the metric has not before been measured, it is the first reliable measurement of structure. Moreover, high resolution image simulation does not work for quasicrystals. HREM images are due to columns of atoms and are not properly interpretable without simulation [xxxiii]. There is no software on the web or elsewhere that can simulate quasicrystals because of their peculiar diffraction. By contrast, every atom on figure 4 and figure 10 is specified, located and measured for size. No other work has achieved any of these. Notice that high resolution electron microscope imaging requires diffraction beam selection and image reconstruction. It is because of the column approximation and phase problem, neither methods nor programs are available for simulating quasicrystals. That is why optimum defocus is more revealing. Atomic resolution HREM was no more successful in

---

[xxxiii] Hirsch, P.; Howie, A.; Nicholson, R.B.; Pashley, D.W.; Whelan, M.J., *Electron Microscopy of thin films*, 2nd ed. 1977, Krieger, NY, especially appendix 5.

simulating atoms in fused silica. Referees should not cite old canards that they neither know about nor understand, piling myth on myth. It is established that our atomic measurement is the first.

Paragraph 3: The prime scientific criteria are evidence and logic; not whether I have referenced Referee's work. The calculations and data are repeatable. I cannot say anything for the kind reviewer's "loss" if he cannot repeat them, except that enlightenment comes from reading and understanding.

3.6 Review 6

3.6.1. Report

"This manuscript is very much in the spirit of works that were published during the first few months of quasicrystals, but were becoming obsolete even then as a better understanding developed. The first line of the abstract is already in error: if Q and Q' are Bragg wavevectors of a quasicrystal diffraction pattern, then Q+Q' also is. Therefore, peaks do include sequences of the form mQ in the reciprocal lattice, where m=1,2,3,... Indeed, most of these peaks have very small amplitudes, leading to the characteristic patterns. But inspection of the patterns shows that, going to larger Q's, the strong peaks do NOT just increase in geometric order; rather, they are distributed with an approximately uniform density in space just like for crystals (I am of course neglecting form factors in this statement which will cause a falloff at sufficiently large wavevector.)

It is universally recognized that quasicrystal Bragg vectors have the form

$$h_1 G_1 + h_2 G_2 + h_3 G_3 + ... + h_D G_D, \quad (*)$$

where h_i are the indices and G_i are the basic recip. lattice vectors. The special property of quasicrystals (which they have in common with the more general class of incommensurate crystals) is that $D > d$ where d is the physical dimension of space, and the G_D's are linearly independent over integers. Indeed, having a Fourier transform of this form is the mathematical definition of "quasiperiodic" which gave its name to "quasicrystals" (short for "quasiperiodic crystal").

"This article appears to be contravening and/or ignoring this idea -- it does not even cite the basic references -- and hence cannot be accepted as legitimate science. Although I should note that Figure 6 -- taking arbitrary sums ofone vector ($h_1 G_1 + h_2 G_2$) from one crystal diffraction pattern and another vector (h_3 G_3 + h_4 G_4) from a second crystal diffraction pattern with cell constants incommensurate to the first -- is mathematically equivalent to what I wrote in (*).

I am sorry that I can't point to easily accessible, pedagogical references about this. Most of them appeared as book chapters, e.g. P. Bak and A. I. Goldman in "Introduction to Quasicrystals" ed. M. V. Jaric (1988). I will point out V. Elser, Phys. Rev. B 32, 4892 (1985) and Acta Crystallogr. A42, 36 (1986); M. DUNEAU and A. KATZ Phys Rev Lett 54, 2688-2691 (cited over 500 times); Also, the idea about the form (*) in general incommensurate crystals has been written about by Ted Janssen, and in a review by A. Yamamoto who has done so much experimentally in structure refinement of incommensurate crystals and of quasicrystals.

"Finally, if the author is arguing for an approach based on treating quasicrystals as a deformation or modulation of a regular lattice, that idea has also been published more than once; I think the first time was M. KURIYAMA and G.G. LONG, Phys Rev Lett 55, 849-851 (1985). [I consider that paper to be a bit pathological even for its time; it got cited only 30 times over the years, the one being 1994]"

3.6.2 Rebuttal

A false hypothesis leads only to no conclusion. The wave vectors are actually quasi-Bragg wave vectors and they add vectorially as in the method of indexation used, and as is typical in electron microscopy. But the high Bragg orders are generally absent, like forbidden lines in X-ray diffraction. That is why the series are often called Fibonacci (review 3, p.82), though the description is inaccurate. As he chooses to debate these facts, it is difficult to know what he has either understood or observed. He is saying that the pattern is not even Fibonacci. I doubt this has ever before been written and it is absurd. Moreover, anyone can see that the 2-fold axial pattern is denser than the 3-fold and 5-fold axial patterns, so his claim of similar densities in crystals and quasicrystals cannot be right. Moreover his claim is flatly contradicted by clear evidence ([6], fig. 2). This contains the two diffraction patterns semi-superposed.

His opinion regarding figure 6 is false: as is evident from the indexation. The two patterns, crossed and diagonal, use the same quasi-lattice constants and are not incommensurate. The two patterns facilitate indexation. We index *ab initio* and we measure the metric for the first time, consistently with the quasi-lattice parameter and with imaging. It is true of course, that he is weakly right that lines are less strongly forbidden in quasicrystals than in face centered cubic or body centered cubic crystals (figure 7b cf. ch. 2 [20]). The trouble is that the metric is not measured from weak lines, and they are not useful in determining structural defects. However the more fundamental trouble is that on his view, the quasi lattice is a continuum ($a_i+b_i\tau$, i=1..3 for three dimensions and a and b positive or negative integers $-\infty < a,b < \infty$) and not a lattice at all. On that view, there is neither a reciprocal quasi-lattice nor a diffraction pattern. Form factors cannot be ignored. It is moreover unnecessary complication to reference incommensurate crystals since a much simpler, fuller (with metric), and conventional analysis is given here. You can tell where he comes from. I was one of the 500 that referenced his citation. The number is *ex judicio* and has no significance now: it was only available at the time. For additional references, he could consult my earlier papers, as cited, but they will not cure his aging "pathology" because physics is not a branch of medicine.

His final paragraph starts with another false hypothesis. I use the unit cell and metric to prove and understand not only "Where the atoms are", but also "Why they are there?" Defensive, aren't they?

Referee's notion of 'legitimate science' is unphilosophic and a logical fallacy (see section 3). If his own work is unreferenced, it is irrelevant, as mentioned in the text. Happily, not all referees take his invincibly ignorant view. As cited earlier (p.81), the most iconic paper in physics has no references.

3.7 Conclusions

3.7.1 Via positiva

3.7.1.1. The Quasi-Bragg law transforms the diffracted geometric series to a hierarchic structure, (where the Bragg law transforms a crystal lattice to its reciprocal lattice).

3.7.1.2 τ is to the icosahedron (length/side) as π is to the sphere (circumference/diameter).

3.7.1.3 The metric is measured, for the first time. It provides in consequence the first atomic measurement in bulk quasicrystals.

3.7.1.4 The simplest indexation is 3-dimensional:
[100] and [111] are same as in the (matrix) fcc lattice;
$[0\bar{1}\tau]$ corresponds to $[0\bar{2}3]$ in fcc;
The stereogram is written 3-D in powers of τ .
3.7.1.5 Quasi-Bragg structure factors match data.
3.7.1.6 All atoms in an Al_6Mn specimen are consistently located and measured.
3.7.1.7 The Driving force is explained.

3.7.2 Via negativa

3.7.2.1 Shechtman's data are not icosahedral.
3.7.2.2 'Bragg diffraction' in 'Fibonacci' series is a contradiction in terms.
3.7.2.3 The diffraction is not Bragg:
The diffraction series are geometric;
higher (arithmetic) Bragg orders are generally forbidden;
all structure factors are zero in Bragg diffraction;
Simulations show the only Bragg order allowed is in second order;
with 6-D indexation no atomic measurements have yet been made.
3.7.2.4 The diffraction is not (unrestricted) Fibonacci;
the ratio between consecutive terms is constant.
3.7.2.5. Interplanar spacings can be measured only with a special metric.
3.7.2.6 Statistics: of 7 referees, 1/3rd voted yes; 1/3rd voted nothing new; 1/3rd voted too new to be true. The alleged broken link of the last type of reviewing is contradicted by alleged repetition in the second type. The fallacy then is that *all* the links (chapter 2 table VI) are supposed repetitive.
3.7.2.7 Completeness requires answers: "Where are the atoms?" "Why are they there?" "Why do they not form metallic glass?"
3.7.2.8 In a scientific revolution, the refereeing system does not work because most reviewers are competitors, while their opinions are tallied without logic:
Extreme diversity in refereeing is a sign of confusion.
Repetition of error makes it orthodox; evidence and logic make science.

3.7.2.9 Scientific journals are political when they depart from the enlightenment of Aristotle and Locke. As IUCr must reject, but with no proper argument, *this is the space to watch.* Of course one has to be unpopular.

3.8 Epitaph *for Linus Pauling and his twin, IUCr:*

"They thought they understood electron microscopy."

Index

2-fold, 27, 28, 49
2-fold axis, 44
3-fold, 26, 47
3-fold axis, 44
5-fold, 24, 45, 46
5-fold axis, 44
Acta Crystallographica A, 1, 6
ad judicium, 72, 73, 80
ad populum, 73, 80, 81
alignment, 62
approximant, 53
Aristotle, 72
atomic map, 23
atomic model, 16
atomic size, 18
atomic sizes, 58
axial, 23
base, 19
Bragg diffraction, 2
Bragg law, 38, 39
Bragg orders, 23
Cartesian axes, 16
chemical species, 31
cluster, 15
compromise spacing effect, 33
concave quad, 16, 17
conventions, 54
convergent-beam electron diffraction, 56
coordination, 17
CSE, 38, 41
defects, 53
density, 18, 19
edge sharing, 15
electron microscopy, 28
Fibonacci, 3, 13, 36, 37
filtering effect, 38
geometric, 13, 22, 23, 36
geometric, 37
golden triad, 16
harmonics, 33, 35
hierarchic, 14
hierarchic method, 55, 61
hole, 32
holes, 53
hopping site, 32
hyperspace, 22
indexation, 19, 23, 43, 51
Indexation, 28, 49
informal fallacies, 72
intensity rankings, 44
interplanar spacing, 33
Interplanar spacings, 21
IUCr, 2
Locke, 72
logarithmic reciprocal space, 13
logarithmically periodic solid, 14
logic, 72
Materials, 6
metric, 14, 33
optimum defocus, 11
Pauling, 12
planar quad, 16
planar quad, 17
quasi Bragg angle, 33
quasi Miller indices, 31, 32
quasi unit cell, 30
quasi Bragg law, 14, 38, 39
scattered intensities, 33
scattering amplitudes, 31
scattering power, 34

second Bragg order, 33
shared atoms, 32
Solid State Communications, 2
stereographic projection, 20
stoichiometry, 17, 59
stretching factor, 17
structural concept, 16
structure factor, 30
subcluster, 15
supercluster, 23
TEM, 52
The International Union of Crystallography, 2
third bright ring, 22
triple point, 16
unit of length, 16
uses, 7
X-ray diffraction, 28

Reference

1. Bursill, L.A.; Peng, J.L., Penrose tiling observed in a quasi-crystal, Nature, **1985** 316, 50-51.
2. Shechtman, D.; Blech, I.; Gratias, D.; Cahn, J.W., Metallic phase with long-range orientational order and no translational symmetry, Phys. Rev. Lett., **1984**, 53, 1951-3.
3. Steurer, W., Twenty years of structure research on quasicrystals. Part I. Pentagonal, octagonal, decagonal and dodecagonoal quasicrystals, Z. Kristallogr. **2004**, 219 391-446.
4. Cervellino, A.; Hailbach, T.; Steurer, W., Structure solution of the basic decagonal Al-Co-Ni phase by the atomic surfaces mod eling method. Acta Cryst. B, **2002** B58, 8-33.
5. Bourdillon, A.J., *Quasicrystals and quasi drivers*, UHRL, USA, 2009. An introduction can be viewed on http://www.youtube.com/watch?v=0LDS0sQQvpk , *i.e.* Youtube/the quasi Bragg law and metric in quasicrystals. A debate is recorded on www.quasicrystal.us .
6. Bourdillon, A.J., Fine line structure in convergent-beam electron diffraction of icosahedral Al6Mn, Phil. Mag. Lett. **1987**, 55, 21-26.
7. Bourdillon, A. J., Nearly free electron band structures in a logarithmically periodic solid, Sol. State Comm. **2009**, 149, 1221-1225.
8. Bourdillon, A.J., *Quasicrystals' 2D tiles in 3D superclusters,* UHRL, USA, 2010. Ch. 2, available also at http://www.quasicrystaltiling.us, 2010.
9. As reference 8, ch. 5.

10. Pauling, L., Apparent icosahedral symmetry is due to directed multiple twinning of cubic crystals, Letters to Nature **1985**, 317, 512-514.
11. Cahn, J.W., Shechtman, D. and Gratias, D., Indexing of icosahedral quasiperiodic crystals, *J.Mat Res.***1986**, 1 13-26.
12. *Directions in mathematical quaiscrystals*, Eds. Baale, M., and Moody, R.V., CRM Monograph series vol. 13, AMS, USA, 2000.
13. Steurer, W., Deloudi, S., Fascinating quasicrystals, Acta Cryst. A, **2008**, A64, 1-11, section 3.
14. Proc. 11th Int. Conf. on quasicrystals 13-18 June, Sapporo, **2010**, Phil.. Mag. 91 [19-21].
15. Hirsch, P.; Howie, A.; Nicholson, R.B.; Pashley, D.W.; Whelan, M.J., *Electron Microscopy of thin films*, 2nd ed. 1977, Krieger, NY, especially appendix 5.
16. Huntley, H.E.,*The Divine proportion,* Dover, UK, 1970.
17. Bourdillon, A.J., *Logarithmically periodic solids,* Nova science, USA, 2011.
18. Levine, D.; Steinhardt, P.J., Quasicrystals. I. Definition and structure, Phys. Rev. B, **1986**, 34, 596-616.
19. Tanaka, M.; Terauchi, M.; Hiraga, K.; Hirabayashi, M., Ultramicroscopy **1985**, 17, 279-285.
20. Cullity, B.D., *Elements of X-ray Diffraction,* 2nd ed., Addison-Wesley. US, 1978, ch. 4.
21. Buxton, B.F.;Rackman, G.H.;Steeds, J.W.; Convergent beam electron diffraction in deposited films, Proc. 9th Int. Cong. on electron microscopy, Toronto, Vol. 1 pp 188-189. 1978.

22. Takakura, H.; Gomez, C.P.; Yamamoto, A.; De Boisieu, M.; Tsai, A.P., Atomic structure of the binary icosahedral Yb-Cd quasicrystal, *Nature Materials*, **2006**, 61, 58-63.
23. CRC Handbook of Chemistry and Physics, Ed. Lide, D.R., 71st ed. CRC Press, Boca Raton, USA, 1990.
24. Abe, E.; Yan, Y.; Pennycook, S.J., Quasicrystals as cluster aggregates, *Nature Materials*, **2004**, 3 759-767.
25. Zhang, H., Kuo, K.H., Transformation of the two-dimensional decagonal quasicrystal to one-dimensional quasicrystals : A phason strain analysis, *Phys.Rev Bt.* **1990**, 41 3482-3487.